A New Vineyard

THE TALE OF UNLIKELY HEROES
DESTINED TO RESTORE ISRAEL

Audrey Lero

CLAY BRIDGES
PRESS

Table of Contents

This story is for any person who has dreams to grow.

Behold, I am doing a new thing; now it springs forth, do you not perceive it? I will make a way in the wilderness and rivers in the desert.
—Isa. 43:19

SPECIAL THANKS

A special thanks to the God of Abraham, Isaac,
and Jacob. The One who grafted me into His fabulous
narrative. The original author of all mystery, drama, and
comedy. Thank you, Father, for inviting me into your story!
And thank you for new wine!

Thank you to my Israeli friends, the ones
who've taught me about the land and the ones who've
ventured there with me. You know who you are. I
am grateful for your encouragement and camaraderie!

Setting the Stage

A Quick History of Israel in the 20th Century to Set the Stage for Gibeah

In the early 1900s, hundreds of thousands of dispersed Jews started migrating to the Ottoman Empire, which at the time was prominently under Arab control. This movement, called Zionism, was spurred by European Jews escaping persecution and establishing their own ancestral home. The British government used force to claim the land on behalf of the Jews through the Balfour Declaration, which read,

> *His Majesty's Government view with favour the establishment in Palestine of a national home for the Jewish people, and will use their best endeavours to facilitate the achievement of this object, it being clearly understood that nothing shall be done which may prejudice the civil and religious rights of existing non-Jewish communities in Palestine, or the rights and political status enjoyed by Jews in any other country.*[1]

It was World War I that further changed the political geography of the Middle East when the Ottoman Empire was defeated. This allowed the Zionist movement to press further into Palestine and Lebanon, while the former surrounding provinces became Arab nations. The League of Nations divided the former Ottoman territories. By 1922, Syria and Lebanon were given to France, and newly developed Iraq and Palestine were given to England.

In the early 1930s, there was a mass Jewish migration to Palestine as the Nazis gained power in Germany. Approximately 150,000 people entered Palestine legally and thousands more illegally; estimates were that 3 percent of Palestine's population was Jewish.

By 1940, Nazi Germany and its allies had conquered much of Europe. Twenty thousand Zionists enlisted with the British to fight in the war. The Zionists fought to gain complete freedom with the hope to one day govern themselves. The British government took the matter to the United Nations Special Committee on Palestine (UNSCOP) to create a solution for the region.

In 1947, UNSCOP reached a decision to partition Palestine into separate Jewish and Palestinian states. Jewish territory was drawn from the Negev Desert to the coastal plain between Tel Aviv and Haifa as well as parts of northern Galilee. The Palestinian Arabs maintained Gaza, Jaffa, the Arab neighborhoods of Galilee, and the West Bank, which is located west of the Jordan River. In all, 1.2 million Jews would be placed in this new territory, and 600,000 Palestinians would be moved to the West Bank region. Jerusalem, the most contested area, would become an international enclave under the trusteeship of the United Nations. The Zionists, led by David Ben-Gurion, accepted this partition plan; however, the Palestinian Arabs and the surrounding Arab states rejected the proposal. The Arabs believed that they should maintain complete control over Palestine and that the Jews were foreigners who needed to be removed from the land.

On May 14, 1948, Israel declared independence and established the Jewish state in Palestine to be called the State of Israel. It declared it would uphold the full social and political equality of all its citizens without distinction of religion, race, or sex. It would also loyally uphold the principles of the United Nations charter. It promised equal rights to Arab inhabitants of the State of Israel and extended peace to all neighboring Arab nations.

In 1949, the Green Line, established in the Armistice Agreement, was drawn as the demarcation line between Israel and its neighboring nations, receiving its name because it was drawn with a green pen on the Armistice Agreement map. In response, the Palestine Liberation Organization (PLO) formed in 1964 to contest land and water ownership and to govern Palestinian relationships with neighboring Arab nations. Despite the intervention of the PLO, tension grew as displaced parties on both sides struggled to accept their newly forced living arrangements within the contested land.

Rising tensions between both groups led to the Six Day War in June 1967 when Israel launched an airstrike against Egypt, Syria, and Jordan. The war ended with the Israeli army occupying Egypt's Sinai Peninsula, Syria's Golan Heights, and Jordan's West Bank, displacing 250,000 Palestinian Arabs. The United Nations Security Council passed Resolution 242 in response to the Six Day War, emphasizing the "inadmissibility of the acquisition of territory by war and the need to work for a just and lasting peace in which every State in the area can live in security."[2]

After the Six Day War, the territories captured by Israel beyond the Green Line came to be designated East Jerusalem, West Bank, Gaza Strip, Golan Heights, and the Sinai Peninsula. Zionists serving in the State of Israel's Knesset, or Parliament, started the Settlement Committee in an effort to establish many moderate-sized cities stretching from Tel Aviv to Jerusalem, especially in the West Bank. The settlement movement surged in Israel in the 1970s as both Jews and Arabs attempted to claim and establish more territory. A settlement was deemed a cooperative or a group sharing an ideology, religious perspective, or desired lifestyle.

These efforts were initially thwarted by the Yom Kippur War in 1973 when Egypt launched a sudden attack on Israel at the east bank of the Suez Canal. At the same time, Syria also launched an attack on Israel at Golan. The war ended with Israel victorious. After the war, the United Nations Security Council issued a cease-fire agreement called Resolution 338. The Settlement Committee ramped up its efforts and continued forming settlements, some of which grew into thriving cities.

Prologue

Aaron Lipkin smiled eagerly as he gazed through a tiny, dusty window of a dark green Israeli Defense Force (IDF) helicopter. It approached the mountaintop and struggled to find a place to land as the mountain was littered with more boulders and dirt than the aerial map had indicated. Israel's mountains were known for their steep and barren peaks, apt to repel human visitors. The propellers kicked up so much dirt in the line of sight that what was anticipated to be a simple landing intensified quickly into a no-landing situation. The IDF pilot yelled through his mouthpiece to the passengers that the landing would not be possible, and they would need to try again on a day after a heavy rain to help lessen the dirt and debris.

Aaron Lipkin's chest rose with indignation. He had waited much too long for this moment and replied to the pilot in a roar, "We have direct orders from the president to land this helicopter on this mountain this day and build a city. We will be landing, and we will begin. You were born for this, Lieutenant, and were trained to fly by the best air force in all the world. I have faith in you and order you to land now!"

The pilot responded, "With all due respect, sir, I cannot afford to put our lives in danger by landing under such conditions."

"With all due respect, Lieutenant, you cannot afford to not land this aircraft!" Aaron replied. "Our forefathers risked their lives to restore this land. Now it's our turn. We must land."

The helicopter made a sharp hairpin turn and landed with a jolting thud into enemy territory, shaking the bodies and minds of the four passengers. Aaron and Ruth Lipkin disembarked slowly, shielding their eyes and ears from the dust storm with light blue-and-white, linen-blended scarves, and surveyed their new home. The couple assisted the co-pilot with unloading the disheveled cargo of camping supplies, clean water, and government food rations.

After all contents were on the ground, the co-pilot saluted Aaron and rejoined the pilot in the cockpit. They lifted the helicopter in the same manner with which they had dropped it—quickly. As the helicopter departed, a faint woosh, woosh, woosh was the only sound that could be heard through the mountainous terrain of the hills of Samaria. Without the cooling breeze from the swift propellers, Aaron and Ruth coughed uncontrollably. They found their canteens and gulped water to relieve pressure from their dust-filled lungs. They looked at one another and, for the first time in years, laughed jovially. They embraced and stood contentedly, knowing that they had made it. They had made it to the mountain.

Aaron gathered his wife in his arms. Now physically still, though hearts pounding, they smelled the fresh land. There were no fruit trees to produce the familiar smells from their childhoods or sounds of car horns, sirens, and shuffling people as in the bustling city they had just left. Ruth held her memories and braced her delicate weight against Aaron's strong and gentle stature.

As Aaron stood atop the mountain, he pointed to the east and said, "My darling, over there will be the hospital."

Turning slightly, he pointed a little to the left of that space and said, "And over there will be the university."

Pointing a little further left of that space, he said, "And over there will be the industrial park. This is my vision. Everything that the people of Tel Aviv have, the people of Gibeah will have, too. We will make a place for them to raise their families, be educated, and invent and manufacture the finest creations earth has yet to know. Together, we will bring the Jews home."

Chapter One

In the 1930s, Leah left her home in Israel to study at the American University of Beirut in Lebanon. Though her family was large, Leah was the only one of seven siblings who wanted an education and had a different worldview. Her parents obliged and paid for her to travel far from home into Arab territory. It was in Beirut that she encountered races and religions vastly different from the ones she knew. She fell in love with new cultures she was exposed to and new people who practiced them. Being away from home allowed Leah to realize how grateful she was for her upbringing in a Jewish community. She, in turn, shared her culture, including holidays, food, and scriptures, with her new friends at the university.

Adam Lipkin was a hard-working, devout farmer and business-man, just like his father. However, instead of choosing to work in the orange grove like his siblings, Adam fell far from the tree, choosing to pursue an education in the marketplace. Under the direction of his father, Samuel, he focused on produce exportation to sell Lipkin oranges abroad. Standing 6 feet 2 inches tall, Adam had a sturdy frame and a handsome face. Spending the profit from successful exportation, he had a custom suit made of rich Italian fabric in order to look his best on his first business trip to Beirut. It was on trips away from home that Adam's heart opened toward his Jewish brothers and sisters living in other parts of the world. Israel reported riots in the Middle East and parts of Europe, but until these trips, Adam had not yet witnessed how severely the Jews were being discriminated against, segregated, and abused because of their heritage. It was on one memorable trip when he was personally called names in the streets and had his life threatened by Arab men for being a Zionist that Adam's perspective changed forever, and his heart was set aflame for a fight.

One night in Beirut after a long day of negotiating prices and logistics to export fruit across international borders, Adam sat down with his father for a quiet meal of tapas and red wine. He had a boldness that night that he hadn't felt before, and he opened up his heart to Samuel about his frustrations. He shared about persecutions he and others were experiencing. It was this open door that made Samuel feel connected to his son in a deep way, and he knew it was time to share his real work. He proceeded to tell Adam about his underground political affiliation with a Zionist group and how the marketplace served as a front for the group's gatherings. Instead of retiring for the evening after dinner as happened most nights traveling together, Samuel invited Adam to stay and meet some of his friends.

A large Lebanese businessman with dark eyes and a dark beard entered the room and joined them at the table. He was well connected politically and also in the global agricultural marketplace. Though Adam was instructed by Samuel to be a fly on the wall for this meeting, Adam could not hold in his curiosity. He asked his new friend for his thoughts on the anti-Semitism in the streets.

The man replied, "Adam, the anti-Semitism is increasing in Lebanon as well as other parts of the Middle East and Europe, as I'm sure you're aware. There is talk of riots, and there are pockets of corrupt government officials who are increasing in their hatred and power. Your father and I are working with other businessmen to start a fund that will move the Jews in these places safely to Israel.

Consider this your swear-in to secrecy. Will you join our efforts to save the people we love?"

Adam looked slowly to his right and left before speaking to the Lebanese businessman. When he determined they were not being watched, he looked directly at the man and said, "How do I know I can trust you? You are not a Jew like me. How do I know this is not a trap?"

The man glanced at Samuel, who nodded. The man then turned back to Adam. "Adam, though we do not pray to the same god at night as we lay down to sleep or celebrate the same feasts and festivals, we are still family. We are both born of Abraham and live in this land. We can live peaceably together, being neighbors and raising our families to build better governments, businesses, and global relations with America and Asia. I have been friends with your father for a long time, and it is because of his brilliant business acumen and connections that my business survived tough times and is now thriving. Your people are blessed. I want to help you save your people as your father has helped and blessed mine."

Adam, holding back a tear, nodded in agreement, and the men devised a plan to begin moving Jews from Lebanon to Israel. This strategy involved hiring Jews to work for their companies to get them documentation and afford them work trips to Israel. Once in Israel, the Jews could simply stay and start a new life in their true, desired home. At this time, Israel was not yet an official state, but spaces were being made with families in Tel Aviv and Jerusalem to house refugees as they crossed the border.

The next day, Adam attended another secret meeting with his father where they connected with a Jewish businessman named Eli. The three men sat down and were talking over cups of strong kahwa when Leah, Eli's daughter, walked in. The men stood in unison to greet the young woman as she approached their table. Adam was immediately captivated by her beauty and enamored by her rich brown hair and sparkling eyes. She carried her femininity confidently and spoke freely with gentle boldness. That intrigued Adam. He desired to know what she thought of him. He had never felt this way before and struggled to focus on anything beyond asking her on a date for the remainder of their meeting. It was as if the passion he felt for the Jews was only occupying half of his heart. The other half was awakened with the introduction of Leah, and his heart became full for the first time.

Though very confident, it took Adam some time to gain the courage to ask Leah to dinner. Knowing trips to Beirut were getting more dangerous and less frequent, he asked her on their last day in the neighboring country. Leah, also intrigued, said yes. On their first date, the two swapped stories of their love for Israel and their rich history as a people. They collectively valued the great lengths and sacrifices their forefathers had made to return to the land. Leah also told Adam of her love for playing the piano. She enjoyed the way the keys felt on her fingertips. She used the instrument to drown out the day's worries and transport her back to the orange groves she loved so dearly at home.

After hours of talking and laughing, Adam took Leah's hands warmly into his own and said lovingly while looking into her eyes of deep wells, "My dearest Leah, you are the most lovely woman I've ever seen. Please move back home to Israel to be with me and to partner with me in this quest to save our brethren and restore Israel."

Leah replied, "Oh sweet Adam, I feel that my heart is beating out of my chest as I carry the same passion and desire for you and the Zionists, but I am torn because I am the only educated one in my family, and I believe God wants me to finish school and do great things with my education. I must decline your offer. I cannot be with you today, though the offer is tempting."

With that, Adam walked Leah back to her university dorm but never lost hope for this woman to be his wife one day.

Adam wrote Leah often to keep her updated on the movement in Israel and the orange harvest. The two saw each other in Beirut when Adam attended business trips with his father. He made it a point to see Leah every opportunity he could afford.

After graduating, Leah moved home to Israel into a house she shared with her parents, siblings, and grandparents. She had only been home a few days when she was startled by a loud knocking at the door. In the middle of rolling dough for that night's matzo soup, Leah went to the door, apron covered with flour. She flung open the door to find Adam standing on the other side of the threshold and holding fresh flowers.

He swept her into his arms and said, "My dearest Leah, you are home in Israel where you belong! Will you take my hand in marriage? You are the one I've been asking God for my whole life. You are strong and valiant. You are beautiful and feminine. You are progressive and classic. I want us to run a business together, have a family, and bring our Jewish brothers and sisters home. We will do great things together. Marry me!"

Leah, seeing Adam framed with the glow from the brilliant noonday sun, replied, "Adam, I love you. Yes!"

Adam and Leah were surrounded by family who had been eavesdropping and were grinning from ear to ear. The family felt complete. Their Ukrainian Jewish daughter had found her Ukrainian Jewish husband.

Eli and Samuel decided to host an engagement celebration for their children. The party was hosted at the family's opulent house nestled among vast orange groves in Israel. Built a few years before by the family, the house stood three stories tall; however, construction was never fully finished because of a shortage of workers who had been recruited to fight in the intermittent Arab uprisings.

On the first story of the house was an office with access to a basement used for storage of farming equipment for the orange groves and as a shelter in case of another uprising. The second story housed one large kitchen and living area where the family spent most of their time. The kitchen would later be divided into two separate kitchens to accommodate Hannah's traditional kosher cooking. Her daughter Leah adopted more progressive non-kosher cooking in Beirut and refused to spend her energy on upholding religious customs when there were more important thoughts on her mind— thoughts of rescuing Jews.

Kosher food was important to many Jews. It required adhering to Torah guidelines for cleanliness, having a rabbi bless the animals before slaughtering, and never cooking or serving meat and milk together.

The second story also housed a great room and library used for entertaining. The rich wood on the walls had overheard many political and social conversations. The third story of the house was originally designed to have several extra bedrooms and bathrooms for visitors but was never finished. It was a nice accident that doubled as a rooftop deck for hosting parties.

While celebrating the engagement of Leah and Adam on the rooftop deck, Eli poured himself a glass of local Cabernet Sauvignon and asked Samuel to share the story of how their family chose Israel to build their lives.

Samuel smiled at Eli and said, "Of course, my friend. You see, when I was just a boy in Ukraine, I apprenticed at the largest fabric store in all the land. It was a beautiful place that stocked the finest silks from Asia. However, its triumphant suggestion could not have been more opposite of my life at the time. My father died suddenly, and my mother was left a young, poor widow with no savings to support our family. In desperation, I decided to drop out of school to work full time at the store. Over time, the store grew in stature and became the preferred provider of fabric to the king's palace. Our rich fabrics adorned the palace and dressed the king himself. I found myself interacting with the palace staff, and by my late teen years, I was running the store. Now, able to travel back and forth to Europe to oversee the importation of fabric, my life's purpose took shape. It was on those trips that I witnessed life deteriorate for our Jewish brethren.

"Though safe at home in Ukraine, we learned of deep-seated hatred and persecution in parts of Europe, and we knew we could not sit idly by," Samuel continued. "We were seamlessly moving fabric, so why could we not use our channels to rescue our friends? We started hiding Jews in our bulk packages and routing them to Israel where we knew they would be safe. We wanted to help further and decided to buy land and join our brethren in Israel. My lovely wife, Rebecca, and I learned through that process that true love is found when a man is willing to lay down his life for his friends. We sold all we had, cashed in years of hard work and favor, and moved to this very space of dirt on which our father Abraham once walked."

"That's a fascinating story, Samuel," Eli replied. "It is similar to ours where we knew we needed to leave Ukraine and trek to this land. We had connections to the Arab orange grove owners through our global marketplace network. Because we had established friendships, even though it was against the law, we convinced the Arab owners to sell us this land and some of the trees. We worked for and with Arabs to build deeper trust and prosper together like our brothers Ishmael and Isaac who went before us. We are pioneers and will continue to buy land until we own all of Israel. That is our goal."

Eli looked around at everyone gathered on the rooftop and said, "This is all for you. We love you, and you are now our family." He pointed below to the neighboring Arab villages and said, "And this is for them. We love them, and they are our family, too—our cousins."

It was the genuine love and inclusivity Eli felt for all who lived in this region that led to his election as mayor of the town, a traditionally Arab community. His family would become one of the founding families, and he would serve for 30 years with Adam at his side as deputy mayor.

Chapter Two

In the early 1940s, Leah and Adam married and then moved into the orange grove house with her family. About two years later, Leah gave birth to Aaron Lipkin at the hospital in Tel Aviv, changing the landscape of Israel forever. She knew from the moment she held him that he would follow in their pioneer footsteps. Aaron looked curiously around the hospital room and made eye contact with his mother, his father, and the doctor, capturing the sights, smells, and sounds of the people who brought him into the world. Adam met his gaze and said, "Aaron, your name means *lofty, exalted, high mountain.* Your life will be great on the earth."

Aaron had a round face with a moderately sized nose, several freckles, and a handsome smile. He was shorter than most boys his age and chose to wear his dark brown hair coiffed and combed over like a politician to give him the appearance of a little more height. Nicknamed *Shepherd Boy,* Aaron did not allow his height to inhibit his sense of adventure or direction. He stood tall for his beliefs. He was as bold as a lion and did not let anyone see the thoughts and feelings of insecurity underneath the roar. His unintimidating stature and fearless personality gave him the ability to easily befriend others, a trait that would benefit him throughout his life.

Soon after Aaron's third birthday, he became a big brother. Tamar was born, and Aaron's life changed forever. She was the first of many lives Aaron vowed to protect, steward, and build up, and he loved his baby sister with his whole heart.

As they grew, Aaron and Tamar walked to nursery school together each day. When Aaron entered the first grade at a new school, Tamar was afraid to be left alone. She gathered her things for nursery school like all the days before, but when it was time to leave the house, she said, "Aaron, I will miss you today." Sniffling a little, she started to leave on her own.

Aaron said, "Tamar, where are you going? Nothing changes just because I graduated. You are my sister. I will still walk you to nursery school and then go on to my school."

The two Lipkin children grew up together in Moshava, learning from their parents how to manage a farm and business. They spent most days playing in the orange grove with their Arab neighborhood friends. At the age of 10, while working alongside his father, Aaron asked, "Dad, do you think this land can produce other things besides oranges?"

I'm sure it could. What do you have in mind?"

"I want to grow grapes, Dad. The well-groomed and manicured rows of a vineyard with staggered trellises adorning green vines seem inviting."

Adam, caught off guard by the advanced thoughts of his 10-year-old, responded, "And where did you hear about vineyards, Son?"

"Well, Dad," responded Aaron, "I heard about them from you. When you read our family passages from the Torah, I remember ones that say we will inhabit houses we did not build and vineyards we did not plant. I think that is great that we are promised to be gifted some, but honestly, I would like to help God out and start building some of my own."

Adam replied, "You are right about those houses and vineyards, Son. They are coming, but there is a purposeful timing to everything. Vines require the right conditions to grow and several years to develop. I do not think today is the right day for us to start that project, but it is coming. I think a better idea for you today is to clear a small space in the backyard and choose some vegetables and flowers that grow easily in our climate in order to practice using your green thumb."

When Aaron turned 13, it was time to initiate him into public worship like all Jewish boys at that age. Aaron was short and skinny with blushing cheeks and big hair. He still looked very much like a young child, not a growing teenager. His friends and family were excited for this coming-of-age moment. They hoped the bar mitzvah would silence the teasing by his taller peers.

The synagogue that hosted the ceremony kept many Jewish orthodox traditions. The men wore all black with hats and kippas and sat together downstairs. The women were in their traditional reverent wigs and sat together upstairs, holding their little ones. Aaron surveyed the crowd that came to support him and thought, *though seated in separate sections, the Jews were unified by tradition and open hearts in reverence for their God. The God of Abraham, Isaac, and Jacob, is keeping watch over them this day.*

As Aaron sang the songs of old with the crowd, Adam's eyes filled with emotion. He was proud of his son and saw much of himself in Aaron. He knew in his spirit that Aaron had a profound destiny. He could sense a wave of family legacy growing and believed this identity, coupled with innate charisma, would allow Aaron to significantly shape the world. Adam's tears morphed into laughter as the ceremony came to a close, and Tamar led the family in throwing candy at Aaron, a Jewish tradition to signify the beginning of a sweet adult life.

A party followed the ceremony at the Lipkin house. As the sun set, adorning the lush landscape with fresh color to close an unprecedented day, the women carried food and beverages to the guests. To expedite this process, a three-story makeshift elevator was manufactured by the men of the house several days prior to the event and after receiving the RSVPs. One thousand attendees needed to be served. They came from all over Israel to honor the beloved Lipkin family. Aaron led the giant crowd in singing and dancing all night. He enjoyed being connected to the people he loved and wanted all to have a good time and celebrate this moment with him.

When Aaron entered his senior year of high school, he dreamed of graduation day. That would be a day that would make his family proud to see their first child graduate, enter the Israeli Defense Force (IDF), and serve two years as required by the developing State of Israel.

In the midst of anticipation building toward this goal, Aaron did the unthinkable and dropped out of school unexpectedly one chilly fall morning. Unbeknownst to his family and friends, he had been battling internal demons for years that told him he did not have what it takes to successfully make it through school or life. He had learned to act confidently to mask those insecurities and was able to hide behind the routine of school and daily farm life. But as the expectations for his life after school grew, torment grew. Aaron's pride and charisma made it challenging for him to sit still in school. His teachers verbally abused him in an attempt to wield his behavior that was perceived as disrespectful. Aaron was pushed to his limits by trusted superiors and ultimately lost faith in his academic abilities, deciding it would be easier to quit and withdraw. Though he was close to his family and friends, the feeling of shame and a sense of failure prevailed, causing him to leave school and not tell a soul.

Sensing a change in Aaron's demeanor from being upbeat and the life of the party to becoming downcast and quiet, Leah waited to address her beloved son a few days later in the non-kosher kitchen. Aaron was in a somber mood, kneading the dough and helping his mother prepare the meal, when he asked to be excused for an early bedtime to study. Leah gently grabbed the sides of her son's round face and tenderly said, "Aaron, I love you. I do not know what you are thinking, but you have a great life ahead of you, and your father and I will do whatever we can to help you. You are not alone."

Aaron cried himself to sleep that night and wished things had been different. He knew he was stronger than what he portrayed that day and felt guilty and awful for lying to his family. *Who am I?* he asked himself. He had never felt as misunderstood and without any direction as he did that day.

Aaron awoke the next morning before sunrise and told his parents he was on his way to his new job at the orange packing plant. He told them he had dropped out of school and would be working for pennies, and that's what he deserved.

His mother kissed his forehead and said, "Aaron, I love you, but that is not what you deserve."

Adam did not make eye contact with Aaron and trudged away.

After three days of not speaking to his son, Adam met Aaron at the orange packing plant early in the morning before his shift began. He asked Aaron to take a walk. The two walked in silence for a while through the trees as the sun began to rise, its rays shining vibrantly against the autumn-colored leaves.

Adam put his arm around Aaron and said, "Son, you are destined for greatness. Your grandfather purchased this land and tilled this soil with his own hands. He partnered with an unlikely people to sustain it and has prepared a place for our people, God's chosen, to be safe and thrive. I understand we are not as religious as our friends who keep kosher and attend the synagogue, but let me tell you who you are, Aaron. You are like Caleb of the Torah. From the moment you were born, you have had a fearlessness about you that allows you to see grapes when others see giants. You are destined to bring the Jews into the Promised Land. Our people who have been dispersed, living all over the globe, are waiting for you to step into who you are and guide them home. There are instructions for you in the books of

Isaiah and Jeremiah found in the Torah that will guide you further. For instance, Jeremiah 31:2–5 reads:

> *Thus says the Lord: "The people who survived the sword found grace in the wilderness; when Israel sought for rest, the Lord appeared to him from far away. I have loved you with an everlasting love; therefore I have continued my faithfulness to you. Again, I will build you, and you shall be built, O virgin Israel! Again you shall adorn yourself with tambourines and shall go forth in the dance of the merrymakers. Again you shall plant vineyards on the mountains of Samaria; the planters shall plant and shall enjoy the fruit."*

This is what your grandparents believed and thus moved to Israel to be part of God's narrative to set the captives free. I caught His vision and married your mother to continue the legacy. Now it's your turn, my son. It's your turn. Bring the Jews home, Aaron. Plant new vineyards."

The two men, now having circled the orange plant several times, embraced, exchanged smiles, and parted. Aaron sat in the dew-covered grass and pondered his father's words. After some time, he picked up his orange crate, placed it at the entrance to the plant, and turned in his notice. He walked home and devised a plan to finish school and join the IDF.

The next day, he said under his breath as he walked back to school, "I am Aaron Lipkin, son of Adam Lipkin, and I will replant the vineyards and bring the Jews home." Using his innate charisma, now reignited, Aaron convinced the school to allow him to take final exams and graduate from high school without finishing in the classroom.

Aaron Lipkin, the graduate, was placed in the communications department as a sergeant of the IDF. He learned Morse code quickly and relayed messages to other IDF officers and allies around the globe. Knowing Aaron's high competency in Arabic, which he had learned from his Moshava neighbors, the IDF sent Aaron to Eilat, Israel's southernmost port at the northern tip of the Red Sea, to translate messages from drug smugglers.

After several months in the Eilat, Aaron was released for a long weekend home. On this trip to the house in Moshava, he rode on a bus with other IDF soldiers where they dodged bullets from Egyptian and Jordanian shooters. He was exhausted from his travel and yearned to sleep in a place where he felt safe. When Aaron appeared at the front door of the house, he startled his family by looking a bit worn and rugged. For the first time in his life, he had long hair both on his face and his head. His skin was bronze, and his stature was more muscular from IDF strength training. His tired arms and calloused hands gruffly embraced each family member. Tamar made him a cup of hot tea and asked him to share stories from the south.

Aaron promised stories after a long nap, to which Leah said, "We love you so much and missed you. We are proud of you each and every day, Aaron!" Aaron beamed with pride at his mother's words, and the warmth of his family rejuvenated his spirit.

After much-needed rest and soaking in a hot bath, Aaron took a walk around the orange grove. While sauntering through his usual lap, he could not help but think how different this farmland was from the desert conditions he had just left. He pondered how to change the landscape of Eilat to look more like his home. He uprooted a cactus from his garden and collected pigeon droppings and fertilizer. He put them into an empty orange sack and set it outside the house to take back on his return to Eilat.

As Aaron prepared to leave home and reenter the Negev hostile territory, he kissed his parents and Tamar good-bye.

As his father dropped him off at the bus stop, he gave him a bear hug and said, "Son, be strong. You are courageous. Do not be afraid of them. We are all so proud of you."

Aaron boarded the bus, and his father waved and smiled until the bus was out of sight. Then Adam wept. He sowed more tears into the ground of Moshava than he ever remembered before. He saw so much of himself in his firstborn and thanked God for such a blessed life.

Upon arriving at the IDF base, Aaron promptly planted his pet cactus in the harsh terrain of Eilat. Not only did the cactus survive, but it thrived in the desert conditions. It grew to its full potential, and its fruit produced seed for others to sprout forth. Today, Eilat is full of cacti holding living water in a desert land. It was here that Aaron recalled the green thumb of his youth and that it extended past the borders of the orange grove. It was also where he recognized his passion to bring life to seemingly dead places.

In the late 1960s, Ruth graduated from high school and joined the IDF for her national obligatory two years of service. She was a Jewish immigrant from Italy whose family escaped Europe's anti-Semitic oppression before World War II. She was strong and beautiful, wise and favored. Forced to grow up quickly when her parents divorced in her teen years, she was self-sufficient. She was estranged from her father and, consequently, her paternal grandparents. She longed for the comfort and affection she saw in committed families.

At age 18, Ruth and Tamar met in the IDF. They quickly became friends and trusted sisters. They shared many meals, many adventures, and many laughs. On weekends off, the duo chose to either travel to Ruth's mother's home in Tel Aviv to party in the city or to the Lipkin house in Moshava to have quiet time on the farm.

On Ruth's first trip to Tamar's home, she fell in love with the Lipkin family. She observed the loud and loving family and yearned for a large, boisterous family of her own one day. There were many aunts, uncles, and cousins who lived close to the Lipkin house and were always eager to gather for dinner and games. The Lipkin family treated Ruth as one of their own and always welcomed her.

Aaron first met Ruth at a Friday night Shabbat dinner at his parents' house when he was 25. He thought she was kind and fun, and he enjoyed her company, but his heart was available for nothing more. His focus at that time was on networking with other Zionists and devising strategies to build a strong Israel. Aaron lived in Tel Aviv and studied political science at the university. His schedule did not allow him to cross paths with Ruth for nearly a year after their initial meeting.

After that year, on a seemingly random Saturday night, Aaron saw Ruth across the room at a friend's party in Tel Aviv. As he locked eyes with her, he felt as if he were seeing her for the first time. Her beauty and familiarity made him want to focus on something other than the Zionist movement for the first time in a long time. The two spent the evening catching up on family and personal matters, and Aaron did not want the evening to end.

The next day, he called his father to share the news and ask him for advice. Adam smiled and said, "Aaron, I am glad you saw Ruth again, but I am warning you to only date Ruth if you're serious. Do not play around with her. She is family."

Aaron and Ruth dated long distance while Aaron lived in Tel Aviv. At that time, he worked for the Israeli Military Intelligence, and Ruth studied zoology at the university in Jerusalem. She shared an apartment with a friend from high school who also had a serious boyfriend, and the four would spend weekends together exploring the Old City.

One quiet night in Jerusalem, not unlike many of their evenings together, Aaron and Ruth finished dinner and walked to the Western Wall. As they strolled through the city, Aaron suddenly stopped in the middle of the Lions' Gate and got down on one knee. Though charismatic and charming, Aaron was not particularly romantic. Ruth did not seem to mind. He asked the question she had been longing to hear: "My beloved Ruth, I love you with my whole heart and want you to be my wife. I cannot imagine my life without you. Will you marry me?" From his pocket, Aaron pulled out a small and unassuming diamond ring that he had bought with his savings at a jewelry store in Tel Aviv.

Ruth was a great match for Aaron and simply loved being in close proximity to him and sharing daily life. As she heard the question, she felt her heart pounding outside of her chest. She responded, "Aaron, my Aaron! Yes, I will marry you!" As simple as that, the two were engaged. They returned to Ruth's apartment, opened a bottle of champagne, and called their friends and families to share the news.

Ruth always felt loved by the Lipkin family but never so loved as the day she arrived to her own wedding, having all the arrangements made on her behalf. Leah and Tamar drove her to the beautiful venue in Tel Aviv. They helped her into a gorgeous white gown that fit her slender frame. They prepared her for the ceremony and party to follow.

Overwhelmed, Ruth said, "Leah, thank you so much for everything. I know it's customary for the families to split the bill for a wedding, so please tell me what this venue, dress, flowers, rabbi, food, drinks, and band cost, and I will tell my mother to reimburse you for our portion."

Leah laughed and gave Ruth a hug. "Ruth, you and your family owe us nothing. We are overjoyed to receive you officially into our family through covenant today. I viewed you as my second daughter from the day I met you, and know God brought us together. This wedding is our gift to you and our way of showing you how much we love you. You are a Lipkin. Just receive from Adam and me today. The gift is free. You are lovely and worth it. I could not have chosen a better match under heaven for my Aaron."

After being wed, Aaron and Ruth Lipkin needed a place to call their own. They heard of a new municipal project on the outskirts of Tel Aviv being built to offer young couples inexpensive housing.

Apartments were selected on the lottery system and required a moderate down payment. The newlyweds were very poor, and Ruth called her mother to ask for advice.

"Aaron and I would like one of these new apartments. Do you have any ideas how we can earn the money quickly?"

"Ruth, I will take your need to the Italian community immediately and raise the funds for you and Aaron," her mother replied. "They love helping and will gladly give where they can. This is what we do. You can pay everyone back when you have the money to do so."

Within a day of making the phone call, Ruth handed Aaron money for a down payment and sent him off to the lottery. At the lottery in Tel Aviv, Aaron watched as two glass bowls sitting on a large table were filled with small white slips of paper. A prominent government official stood on a stage and grabbed a name out of the bowl on the left and an apartment number out of the bowl on the right. Each name and number combination was called. If the person who was announced was present with the down payment in hand, that person would be granted an apartment that day. Aaron went home to Ruth that night with a new apartment key and the flower arrangement from the head table where he had sat while watching the process.

They moved into their first apartment after spending their first year of marriage together in the Moshava house. Ruth, eight months pregnant with their first daughter, was ready for the space. It was small with two bedrooms on the first floor and adjacent to a neighborhood where several government officials lived and were also working to establish the new country of Israel. Men who would later become high-ranking officials in the Israeli government lived nearby. They were not in favor of the inexpensive and seemingly provincial governmental housing project, but little did they know the neighborhood they abhorred would house the neighbors they would need to achieve their goals.

Aaron started attending meetings held by his neighbors and engaging in more complex Zionist plans. A group called the Settlement Committee formed from these discussions. Under the directive of the president of Israel, this group consisted of men and women who were willing to build cities in desolate places. After several months of these meetings, the couple arrived home late one night. It was a night that Ruth would never forget. She froze in a trance-like state and looked around her cozy apartment. She knew in her spirit to take in the sights and smells of this decadent comfort. She had a feeling they would not be living there much longer.

Chapter Three

Approaching 1980, just after the Yom Kippur War, Aaron presented a plan to the Settlement Committee. Surrounding him at the conference table were Jewish men and women he had met and established deep relationships with from their time together in Tel Aviv.

"Ladies and gentlemen, I am here to evoke passion for our goal," Aaron said. "We have worked together for years to craft Zionist growth strategies. Our goal is to find a spot for a nucleus upon which we can grow other small cities, ultimately creating an unbreakable chain from Tel Aviv to Jerusalem. That will unify our people and make attacks much harder for our enemies. Through years of friendship, I have earned your trust and respect as a pioneer for our country. You've entrusted me to lead a team and start a new city. I have prayed and surveyed the land. I am ready to present my findings to you and the land on which I have chosen to build."

The Zionists at the conference table, warm blood pumping at the thought of charting new territory, cheered in festive agreement with their fearless comrade.

The Settlement Committee was comprised of 12 Zionist members. They all had military backgrounds, Jewish genes, and strong convictions regarding the land of Israel. They felt it was their duty to establish settlements and have a physical presence in strategic areas to reclaim territory that was taken from their ancestors over a thousand years before. These men and women empowered IDF graduates and officers to volunteer to start settlements. They encouraged them that they were on the earth for this time and had the right training and skills to create new towns. The Settlement Committee determined each settlement would start with 40 families in order to establish a manageable number of people, yet large enough to feel like a kibbutz.

Aaron and a few of his close friends had scouted the land for the ideal location in Samaria. They searched within a 40-kilometer radius of their anchor city, Tel Aviv, knowing that the law required businesses to provide transportation for their employees up to that distance.

During the search, three locations were highlighted. The first was near a large farm teeming with resources and agriculture. The second was near the ocean, which provided a natural border, shielding them from ground attacks and offering access to seafood and transportation. The third choice was a place called Gibeah, named by its Arab neighbors who claimed that nothing could grow on this mountain. Some even called it the Mountain of Death due to its unforgiving and impossible-to-navigate terrain. It was a dead place, unfit for living, growing, and production as it lost its Jewish inhabitants in the Diaspora shortly after the crucifixion of Jesus. Though seemingly barren and forgotten, Aaron saw a future there. It was exactly 40 kilometers east of Tel Aviv and 40 kilometers west of Jordan. Chuckling to himself, Aaron fancied the spot like a nose, centrally positioned in Israel.

The Settlement Committee appreciated Aaron's enthusiasm to launch his city in Gibeah but pushed back strongly on its location as a place to build. Aaron met with the members individually to understand their concerns regarding a lack of natural resources and its position in the middle of Arab territory. During the meetings, he assured each member that the plan would work, convincing them through his unshakable peace. He told them how his spirit came alive when he walked and prayed on the land. What moved the committee's vote to unanimous was Aaron's prophetic speech, which they could not deny.

"In the Torah," he began, "40 is often used for time periods—40 days or 40 years, two distinct epochs. Rain fell for 40 days and 40 nights during the Great Flood.[1] Spies explored the Promised Land for 40 days.[2] The Hebrews lived in the Sinai Desert for 40 years.[3] Moses's 120-year life was divided into 40-year segments separating his initial education in Egypt, his time alone in the wilderness, and his leading the Jews out of slavery to the door of their promise.[4] Gibeah is 40 kilometers in each direction to our neighboring borders. It is a mountaintop perfectly situated to host 40 brave families."

In the Torah, when Moses sent spies to the Promised Land to see if it was, indeed, safe to enter, only Caleb and Joshua returned, saying that the grapes were the largest they had ever seen and forgetting to mention the armies of giant men. The others who were sent spoke only of the seemingly insurmountable giants and made no mention of the grapes at all. Caleb and Joshua trusted that God brought them to this land to enjoy the grapes, among other abundances, and did not fear its current inhabitants.

Like Caleb, Aaron saw vineyards at Gibeah when others saw boulders. He trusted that the peace he felt about this place was from God, and he was determined to bring the Mountain of Death back to life. Now in his mid-30s, Aaron started to prepare a place to bring the Jews home.

Aaron arrived home that evening after a long meeting. He greeted Ruth with a long face and hunched shoulders. She ran to the door to embrace him, and then he cracked a smile that resembled the parting of the Red Sea and shared the good news with her. He said, "Darling, pack your bags. We are moving to Gibeah. We are officially pioneers."

Ruth smiled back and slapped him on the arm, responding, "Aaron, you have always been full of surprises! We will do this together. Now is the time."

The couple had never been in a helicopter before. They were slightly embarrassed and did not want to share with the others that they felt quite nervous and a little queasy in anticipation of the journey ahead. As the helicopter departed from Tel Aviv, the couple waved to their former home and all their friends. They waved good-bye to the previous chapter of their lives and looked forward, ready to embrace the next one.

Chapter Four

A few days after their helicopter landed in Gibeah, Aaron and Ruth were joined by their two children and 39 other families. These men, women, and children went to meet the Lipkins in an IDF military convoy of trucks and tanks filled with government-issued tents, food, and basic home furnishings for survival. The Settlement Committee allotted a certain amount of rations and finances to each of Israel's new settlements.

Shielding his face from the hot sun, Aaron sprinted down the mountain to welcome the new settlers. He helped them unload their minimal belongings and worked tirelessly to pitch tents so everyone had a place to camp before nightfall. As Aaron worked with his hands to cultivate this place for habitation, he remembered the significance of why they chose this land.

There is a story in the Torah of a man named Jacob; every Jewish boy knows the story. Jacob was the son of Isaac and grandson of father Abraham. From Jacob's family line would come the 12 tribes of Israel and all the Jews, God's chosen people. God gave Abraham a promise that his descendants would be as vast as the stars when he looked up at night. The only problem was that this promise was given to Abraham when he was nearly 100 years old and did not have any biological children. He and his wife, Sarah, laughed at God's idea and struggled to believe the promise would actually come to pass.

As Aaron looked down at his dirt-covered boots atop the freshly inhabited Gibeah, he dreamed about this Jewish tale and became more aware of its reality and the depth of faith it took for Abraham and Sarah to believe. Their grandson, Jacob, one of the promised children, left home to search for his own way in life. Through counsel from his mother, Jacob went to the very place Abraham had received the promise from God. It was there that Jacob found a well dug by his grandfather and stopped for a drink of water. At the well, he saw the most beautiful woman he'd ever laid eyes on and instantly chose her as his bride. Her name was Rachel. However, Jacob had to wait a very long time to marry Rachel and endured intense trials that tested his faith. Nonetheless, he tirelessly persisted and eventually married the woman he met at the well. Through this union, the Jewish world was populated, and it was at this well that more of God's promises were fulfilled.

In that sobering moment, Aaron was humbled after recounting the great promises fulfilled near where he physically stood. He began to pray and ask God to grant him favor to fulfill his destiny to bring the Jews back home. Aaron had a vision. He wanted to fulfill the promises of God and thought there was no better place to do it than the place where it all began.

As the sun was setting, Aaron gathered all 40 families around a blazing campfire to reiterate his vision for the land. He unrolled a large Israeli flag and thrust its base into the earth so its white and blue could wave proudly in the wind. He shared the message about Jacob to remind these people who they were. They were descendants of Abraham, Isaac, and Jacob, and they served the same God. It was this God who brought them to this place on this day to continue subduing the earth and multiplying for the advancement of His kingdom. Aaron told them they were chosen for this mission and had all the provision and resources they needed to complete it. He thanked them again for their willingness to come, and then he made a joke, declaring them officially the 13th tribe.

"After we recruited all of you, my Zionist friends, you endured rigorous interviews to test your ideology and motivation for pioneering a new trail. We challenged you in group conversations and spontaneous home visits to see the purity of your lifestyles and commitment to family and country. You were chosen because the Settlement Committee believes you are more than capable of living together in harmony and will cooperate and connect with the tough environment. We believe you have mental and emotional tenacity, a good work ethic, and the willingness it will take to create a new life here. My beautiful wife, Ruth, has named you the people of patience."

Having their full attention, he asked them to stand and look ahead as he pointed across the pink, blue, and yellow horizon. "Over there will be the hospital." Turning slightly, he pointed a little to the left of that space and said, "And over there will be the university."

Pointing a little further left of that space, he said, "And over there will be the synagogue. This is my vision. Everything that the people of Tel Aviv have, the people of Gibeah will have, too. We will make a place for you and others to come and have their babies, be educated, and worship our God. Together, we will bring the Jews home."

With that, the people went to sleep for their first night under the same stars that covered their forefathers thousands of years before. Aaron joined Ruth in their tent and snuggled under the pallet of blankets she had made for warmth in their temporary home.

Aaron leaned over and grabbed his handheld radio. Pressing the button, he said, "Sir, the families made it safely. We are closing out our second day in Gibeah. Let the record books show we have founded our city. Good night."

As Aaron released the button, he and Ruth heard a confident male voice project through the radio's speaker, "Lipkins, well done.

On behalf of the Settlement Committee, I say you have made our country proud. Good night."

As the next day in Gibeah dawned, Aaron awoke with the excitement of a young boy on the first day of Chanukah. His first order of business was to cultivate a neighborhood. Israeli settlers were used to occupying new land and building a traditional Israeli kibbutz, a compound for protection where many families lived together and shared all food, clothing, money, and other essentials in one space. Aaron's vision included many houses where individual families could live independently yet close enough to their neighbors to keep the relational intimacy of the kibbutz.

Aaron focused next on how to provide food for the community. Scooping up some dry sand and running it through his fingers, he looked out on the vast desert before him and thought, *First, I will need fresh soil.* Aaron radioed into the Settlement Committee and asked for several tons of topsoil and fertilizer to be delivered in order to build a community garden. He also asked for extra water in addition to the rationed portion being delivered for drinking, cooking, and bathing. A voice radioed back, saying that these resources were not in the planned budget and that the request was denied. As the radio went silent, Aaron's heart sank. He realized he would not be able to fully rely on his government like he thought to provide support for Gibeah. Not knowing where to turn for help, he knelt and prayed. From now on, he would need to trust his God wholeheartedly for resources.

That evening, Aaron gathered the people of Gibeah and said, "When you go to your jobs tomorrow in Tel Aviv and Jerusalem, before you pack into the military convoys to come home to Gibeah, fill your bags with all the soil you can find and bring it with you."

As weeks passed, the people collected enough soil to start gardens. The growth of vegetation brought birds, insects, and small animals to the mountain. The growth also brought curious visitors from surrounding Arab villages. Though Aaron was friends with many people of Arab descent in Moshava and spoke Arabic flawlessly, wisdom told him to radio the Settlement Committee and ask for IDF soldiers to stand guard for a preventative measure. His request, however, was denied as money was not allocated for guards at that time.

Frustrated with his government and determined to keep his people safe in the event of an attack, Aaron implemented two courses of action. First, he empowered his people to patrol their own territory by creating a group of night watchmen. All citizens of Gibeah were required to take shifts; only children and pregnant women were exempt. Each shift consisted of five vigilant citizens, lasted four hours, and passed seamlessly from group to group to account for all 24 hours. Then Aaron came down from Gibeah and met his Arab neighbors face to face. He was determined to befriend the leaders of the villages and live in peace in the middle of the state, just like his father had taught him at the orange grove. *They were family, after all*, thought Aaron, and he wanted to create a space for all to thrive.

Military convoys transported people and cargo back and forth between Gibeah and established cities. Roads were naturally developed, and traditional cars and trucks could drive safely on them, though they were not yet paved. As the citizens of Gibeah began driving themselves each day to their jobs, they started grocery shopping to shift some government dependence on food rations. Shopping was done in groups, and a designated shopper drove into Tel Aviv or Jerusalem and bought food for all. Eventually, a family in the community opened a local grocery store to meet the needs of the growing town. Aaron noted that this was the first sign of commerce, and the Gibeah marketplace was born.

After several months of Gibeah being established, the Settlement Committee sent a few generators to the mountain. Houses were connected with operating lines and shared power equally.

One row of houses would get surges of power for an hour at a time. The women collectively decided that Thursdays were baking days. They would gather together in a group and assign ovens and times to each woman. The men roared with laughter as they watched their wives run from house to house every hour as the electricity moved. Thursdays became known affectionately as *the dancing of the ovens.*

As the city and population grew, so did the amount of trash. One morning, Aaron witnessed something remarkable. Two teenage boys were walking from house to house, volunteering to collect the trash for families and piling it near the city to burn. However, the smell of burning trash was more offensive to the people of Gibeah than the original problematic smell of raw trash. Aaron taught the boys how to turn their good deed into a business, and they began to charge money to collect the trash. They used the earnings to drive the trash outside the city to a nearby dump. The second official business in Gibeah was born.

More growth and more commerce kept safety top of mind. Aaron's persistent requests for IDF patrol eventually succeeded. However, the citizens of Gibeah had such pride in their land that they continued to watch in shifts alongside these guards.

Word was getting out about this small, new city, and the Settlement Committee received many inquiries from potential pioneers. With growth came the need for more developed homes, and Aaron was determined to open the borders to more Israelis. Gibeah residents quickly transitioned from temporary tents to concrete dwellings that resembled cargo shipping containers. The people were grateful to move into these government-supplied single-family homes that, though devoid of windows, provided shelter from the elements. Each family received its own dwelling measuring 500 square feet. Bachelors moving into the neighborhood shared a dwelling with at least three bunk beds with other bachelors.

Aaron knew more government-sponsored temporary dwellings would not be the answer for housing and made a trip to Tel Aviv to visit his friends at the government to personally ask for permanent housing. Though they told Aaron numerous times that funding was not available, Aaron requested it in person anyway. He made the four-hour round trip only to come back empty-handed.

Weathered and exhausted, he walked into his temporary dwelling where he found Ruth cooking dinner. She was always waiting for him with open arms and warm food. He smiled and calmed when he smelled the broiling soft white tilapia.

Aaron collapsed into her arms and said, "My beautiful darling, what am I going to do? Gibeah is growing so quickly, but we need permanent homes. Our government has left us out here to fend for ourselves!"

"Aaron, we did not move here to be dependent on our government," Ruth replied. "We moved here to be obedient to our God. He will show you how to make space for the Jews on this mountain. Sit here and rest. I will make your favorite matzo ball soup for dinner, just like your mother used to make to warm your soul."

Ruth got up and put their two children to sleep and turned on the four-legged record player they lugged from Moshava. With gentle jazz playing in the background, the couple spent dinner resting in each other's arms, reminiscing of their days in the house with Adam and Leah. They dreamed of their own future family, teeming with grandchildren and all the adopted ones who sought refuge from the known world within the boundary lines of Gibeah. With a full stomach, Aaron exhaled and knew tomorrow would hold fresh hope and solutions.

Chapter Five

Aaron nearly jumped out of bed the next morning and exclaimed, "Ruth, I've got it. I know how I can get the people of Gibeah into permanent housing!"

"That's great!" Ruth replied, rubbing her eyes. "How?"

Aaron said, "I'm going to build a zoo."

Gently punching Aaron's chest, Ruth said, "Aaron, you always know how to make me laugh. A zoo! That is hilarious. Do you plan to take us back to the garden of Eden so we can live with the animals?"

By this point, Aaron had his right leg into his khaki pants and was shifting weight for the left one to join. Tossing on his pristinely pressed shirt, he pulled her face close and kissed her lips. "I'll tell you the whole vision when I get home tonight."

And he was off, leaving Ruth to wonder how much stress this large new project might put on their growing family.

Aaron ran next door to his best friend David's house to share the good news about how they would solve the permanent housing problem. David and his family moved to Gibeah at Aaron's request to fulfill the need for a local doctor. Aaron asked the Settlement Committee to screen applicants and send doctors to him to interview.

David had been a general practitioner in Tel Aviv. He stood a little taller than Aaron and had thick brown hair and thick-rimmed glasses that made him look the part. David had a great sense of humor. He was passionate about health and had spent his free time educating anyone who would listen about how to increase their lifespan. Like Aaron, he was similarly convicted about helping the Jewish people regain territory and wanted to do his part to grow the nation. During his interview, David shared with Aaron that his own father was a Zionist pioneer. About the same age with similar upbringings and worldviews, the two men became instant friends with a deep soul connection.

Aaron fully trusted David, and Ruth was grateful that David and his family lived next door. Their presence gave Aaron a sense of security when he traveled and had to leave Ruth and his children at home. David's family reciprocated the affection and enjoyed living next door to the Lipkins. It was as if they had always been one large family.

As David's wife poured the two men cups of coffee, Aaron said, "David, here's what we will do. We will build Israel's first zoo and charge admission for people to see the animals. Not only will this attract new people to Gibeah, but it will make money quickly. We will use the proceeds to fund permanent housing. Let's not wait on the government to decide these things for us any longer. Let's have the people of Israel fund the people of Israel."

David smiled at his friend and replied, "Aaron, if this idea had come from anyone else, I would have told them they were crazy, but because it's you, I actually believe it's a fantastic idea!"

Aaron called farmers and ranchers he knew through his father who had access to animals. He shared his vision, and while some people laughed at him, others were willing to support the idea and donate animals from their herds. After a few months of phone calls and persistence, Aaron had 10 animals to start a petting zoo. They included chickens, goats, donkeys, snakes, and a zebra surprisingly named Noah, which became the main attraction.

On opening day of the zoo, Ruth stood next to Aaron, posing for media coverage photos, and said, "And I thought I was the one who had the zoology degree. Again, you surprise me, Mr. Lipkin, with your ability to do anything to bring the Jews home."

A grin spread across Aaron's round face. The zoo immediately increased the visibility of Gibeah and soon paid for itself. It generated so much profit that it was able to put excess cash flow into a real estate fund for permanent housing. The success of the zoo gave Aaron the confidence to explore other ways to grow Gibeah without seeking government assistance. He was learning how to respond to the shifting political landscape in order to get his city built.

Permanent houses that were constructed formed several neighborhoods, all intricately connected to a city center. People were encouraged to purchase homes and let their roots grow deeply into Gibeah soil. If someone could not afford to purchase a home, Aaron arranged for the person to rent to own with the caveat that the person also take financial courses to understand how to finance the purchase and manage money. These new homes were three to four bedrooms in size with extendable options for future planned growth.

With new permanent dwellings and an influx of immigrants on their way, the people of Gibeah needed their own governance. They wanted a leader to represent them to the state and to the world. Though Aaron seemed like the clear choice, there were several other men who also cared greatly for the people and helped shape the city.

Aaron shared his desire to run for mayor with Ruth one night over dinner. Ruth's countenance visibly changed from cheerful to somber when she heard his words. She set her fork down and made eye contact with her beloved. "Aaron, I never dreamed we would do all that we have as we pioneered this city, but I do not want my children or me to be in the public eye beyond what it is today. I have seen what the media did to your grandfather and father over the years, and I do not want that public opinion over our lives here, too."

Aaron took Ruth's hands in his own and replied, "I know this is a big step, but our city is in need of official leadership. We are called by God, Ruth. I believe it is my destiny to lead these people, and I'd like to run with your blessing. The next step in achieving this is to run for office."

Ruth slowly nodded and smiled. "Aaron, once you make a decision, you never back down. I do believe this is part of your destiny. Please just promise me that you will do it differently than your father and love our family as much as you love the people of Gibeah."

Aaron nodded. "I will."

Aaron picked up the phone and called his father to ask for guidance in running a mayoral campaign. Adam responded with great pride. "My boy! I would be honored to help. This is what you were born to do. I knew this phone call would come one day!"

The first thing his father recommended was to hire a campaign manager, preferably a female as he felt they were more organized. He emphasized that this woman would also need to be tough enough to make the hard calls and shake hands with both constituents and diplomats.

"Find someone trustworthy who treats you like family," his father said, "someone in their youth who is filled with the energy to press through long days of work and travel around Israel. Someone who can discern who is privy to information and someone who supports you until the end."

Aaron shared his father's words with Ruth and David, who both helped search for a campaign manager. Using their connections, they were introduced to a young woman who had just moved to Gibeah from Tel Aviv and had previously worked for some members of the Knesset, Israel's highest-ranking governing body.

Diana, wearing a sharp black pantsuit, greeted Aaron and Ruth with a firm handshake at their home for her interview and passed inspection with flying colors. She knew the ins and outs of Israeli politics and was well traveled. She was a perfect blend of shrewd and kind. *She feels like family, like a daughter*, Aaron thought. He asked her to start immediately.

Diana was in her late 20s. She had a medium build, long blonde hair, and blue eyes. She loved her Jewish heritage and was a Zionist to her core. She was single and desired to marry a zealous Zionist man who shared her convictions.

Aaron and Ruth's home became campaign headquarters. Each morning, Aaron, Diana, Ruth, and David met and talked through Aaron's daily schedule. Ruth was Aaron's cheerleader, and David his sidekick. The group created flyers listing all Aaron's accomplishments since founding the city. They reminded the people how they came to the land and the vision for future growth to fully make it home.

The campaign for mayor came during a busy season for Aaron as he was also attending law school and commuting to and from Tel Aviv. In the days leading up to the election, Aaron was getting little sleep and becoming increasingly nervous. He feared something small would inevitably go wrong, as it usually did when Aaron was tackling anything new. He was aware that his charisma was polarizing and prayed to God that there were more in favor of him than opposed.

Also during that time, Ruth was asked to manage the animals while the primary zookeeper was on pregnancy leave. Little did Ruth know that this temporary position would turn into a 20-year career. Both Ruth and Aaron noticed the strain their commitments were placing on their marriage and family but believed these sacrifices would only be temporary.

After several months of campaigning, Aaron waited in anticipation for the media to announce who would be the first mayor of Gibeah. He gathered with friends and family at the only restaurant in the city to watch the results come in on the television. The same anticipation could be felt in Moshava where Aaron's friends and family sat glued to the television in anticipation. If Aaron were to win, he would follow in his father's and grandfather's footsteps, continuing the legacy they started as public servants for their community.

As Aaron looked around at the faces he loved, he remembered the day in the orange grove with his father. He said under his breath, "I am Aaron Lipkin, son of Adam, grandson of Samuel. I am like Caleb of the Torah, God's chosen to replant the vineyards and bring the Jews home."

Just then, news came in that Aaron had won the election by a landslide. Gibeah had its first official mayor. It was a proud day for all involved. Aaron received a flood of phone calls from his IDF buddies whom he had labored alongside for years. As he gave his acceptance speech, Aaron publicly shared his convictions for his people and their land in the contested West Bank, which felt forgotten by most. Just seven years after Aaron and Ruth landed on this mountain, Gibeah was finally on the map.

Chapter Six

The morning after the election, Ruth, now the first lady of Gibeah, awoke to a sunny day and looked out over the city from the second level of her new permanent house. No longer a desert of boulders and dirt, the city was now teeming with life. As Aaron ate breakfast, Ruth called down, "Aaron, we are the people that Ezekiel spoke about. I see it so clearly now."

She picked up the family Torah and read part of Ezekiel 36 aloud.

> *And they will say, "This land that was desolate has become like the garden of Eden, and the waste and desolate and ruined cities are now fortified and inhabited." Then the nations which are left all around you shall know that I am the* Lord;*, I have rebuilt the ruined places and replanted that which was desolate. I am the* Lord; *I have spoken, and I will do it.*
>
> —Ezek. 36:35–36

She paused somberly and then continued reading aloud with a grateful heart in Ezekiel 37.[1] "The mountains of Israel," she exclaimed. "Our inheritance as a nation. Dry bones. Fresh breath of life. Sinews and flesh! It's all there, Aaron! We are those people."

Aaron set his fork down on the table and ascended the staircase to join his wife. As he embraced her, he smiled, his eyes fixed on the city below. "Yes, we are those people. Thank you, God. We will bring your people home."

David stayed a trusted friend and stood next to Aaron as deputy mayor of Gibeah. The new governmental order, international recognition, and the feeling of acceptance by the State of Israel brought a renewed sense of community throughout the city. The camaraderie felt throughout was different from what these settlers had known in the cities from which they came. That made it easy for new pioneers to join and meet others. The people of Gibeah started an open-door policy, partly because their doors did not have locks but mostly to welcome neighbors and guests at all times with a sense of family. These pioneers desired to progress from the kibbutz nature of their ancestors yet found themselves living as a large family community sharing their resources.

As mayor of the city of Gibeah, Aaron turned one of the temporary buildings into an office and started connecting with every governmental official he knew in Israel. With David, his deputy mayor, and Diana, his capable assistant, he set off to make plans to fulfill the vision.

Ninety percent of the first settlers weren't religious, but like Aaron, they had strong convictions that they were assigned to this place by God and should erect a memorial of stones in His honor. They also wanted to thank and worship God and build a synagogue. The first synagogue of Gibeah was built inside one of the permanent houses. As these first families were building the structure, they realized that no one had a traditional Torah scroll, which was required in order for the structure to be deemed an official place of worship. A few wealthy religious Israeli families in Tel Aviv heard of the need and donated money to purchase the expensive treasured artifact. These Israelis believed that in order for Samaria to thrive, it would need a scroll prophesying their very existence.

Chapter Seven

In the mid-1990s, while pondering how to expand his fundraising efforts beyond Israel, Mayor Lipkin was interrupted by a knock on his office door. Diana walked into the office and sat down on a squeaky chair across the desk from Aaron. She said, "Joe is on the phone, and he said it is of utmost importance and would hold until he could speak with you. He sounded almost breathless with excitement! Can I listen in on the call?" With that, Aaron pushed his phone's speaker button, and the two listened.

Joe, a European, Jewish, middle-aged man but Christian by religion, lived in Gibeah in its early days and had since moved to the United States to grow his business internationally. Joe and Aaron had not seen each other in several years, but still there remained a great level of trust between them. Joe had a straight line to Aaron's ear at a moment's notice. While Joe was living in the United States, God put him in the path of an Australian pastor named John who lived in Texas. John held a weekly dinner where men gathered to hear him preach on the goodness of God.

Aaron could hear his friend's voice on the other end vibrantly say, "Aaron, my friend, how I have missed you! You will be astounded by what God is doing for my business in the United States and the kinds of people I am meeting. Have I told you about John? He is a pastor who leads the Bible study I attend. He's radical and teaches us about how Christians are grafted into the lineage of Abraham and are connected to Jews. He said it is important for Israel not to give up any of the land we have rightfully taken on behalf of our people, and that if he were to give up the land, he would fall out of power quickly because he is going against God's heart for His people."

Aaron replied, "Joe, that is quite the story, and this John character sounds just about as crazy as me. How do you know it is true? I hear about these wacky preachers all the time with their God ideas and their Jerusalem complexes. They think they know what is best for the Jews and our land. How do you qualify the legitimacy of this man's claim?"

"Aaron, you never change. That was exactly what I knew you would ask. John didn't share this story with me, only to the men at the Bible study. He told them he asked God to confirm the words he believed he heard, and God said, 'A man will approach you and give you a handwritten note with these words: 'Go to where you need to go.'"

Intrigued, Aaron leaned closer to the phone.

"And," Joe continued, "I ended up being that man! I ran into John later that week at a philanthropic event at an Austin hotel. Not knowing any of this, I approached him and gave him the note as instructed by God. John laughed hysterically and hugged me. He said, 'Now, I am released to share with the prime minister the words from God. Now, how am I to meet the prime minister?'"

Aaron interrupted his friend on the phone and asked, "And that's why you're calling me?"

Joe replied, "Yes, it is, old friend. Yes. It. Is. John is definitely as crazy as you are, and God has given him important words to share with our leader. It is our duty to make sure he is heard."

Aaron told Joe he was skeptical and did not have time to promote a Christian prophet but would ask God for direction on how to proceed. As he ended the call, he looked across the desk at his faithful assistant and said, "Diana, this is the kind of stuff we were made for. I was just sitting here thinking of how I could get across the ocean to start asking Christians to join us in our fight for land, and then God brings them to me."

Diana, perplexed, smiled and nodded. "Aaron," she uttered, "I do not understand why God would go outside our walls to bring us news. If we are His chosen people, shouldn't He be talking directly to us the same way He spoke to our forefathers with burning bushes, clouds, and angels? We all know how Christians treated us during the Crusades. I do not trust them."

But Aaron did not hear his assistant. His mind was already envisioning the possibilities of big checks from Christians to fund Gibeah's growth.

In less than a week, Aaron asked Joe to put him in touch with John. He felt peace about pursuing the Christian for American introductions. Aaron dialed John's international number and said, "John, Joe told me your story and the words that you are carrying from your God for our people. When can you come and share them with me? I would like to hear more about what you believe God told you about our land and our people."

"It is nice to meet you over the phone, Mr. Mayor," John replied. "I, unfortunately, am very busy this month and am not able to travel to Israel and deliver the message to you in person."

John heard silence on his end of the phone followed by a muffled yell. "Diana! Book me a flight to America next week. I'm going to visit my new friend, John."

Ten days after their phone call, John picked Aaron up at the Austin, Texas, airport in his humble white sedan. It was Aaron's first trip to the United States. Aaron greeted John and said in very broken English, "Hello John, I've never met a pastor in real life before. Are you going to pray for me before we begin?" Though fluent in several languages, Aaron had never learned English, but he was determined to improve quickly with new work to do.

John replied in a thick Australian accent. "I've picked up many men of God from this airport, and never has anyone asked me that before. Sure, let's pray."

As John racked his brain on how to pray for this Jewish man, he thought, *Why not go big and pray the sinner's prayer and attempt to save his soul?* "Repeat after me," said John while keeping one eye open and on his new Jewish friend.

"Dear God, I know that I am a sinner and there is nothing that I can do to save myself, even though I've kept the Passover. I confess my complete helplessness to forgive my own sin or to work my way to heaven. At this moment I trust Jesus—I mean Yeshua (that's what you call him, right?)—alone as the One who bore my sin when He died on the cross. I believe that Jesus was raised from the dead on the third day. Thank you that I can face death now that you are my Savior. In Jesus's, I mean Yeshua's name. Amen."

Instead of repeating, Aaron laughed and said, "So that's how you Christians talk to our God."

Feeling sheepish, John quickly recognized his error and joined Aaron in laughter. John said, "I am so embarrassed, sir. I do not know why I said that. I do not know many Jews."

John vowed to himself to learn more about the Jewish people but realized he only needed to treat Aaron like anyone else, a person, a child of God. He drove Aaron to the same hotel where Joe delivered the message and determined there was no better place than back at the well to get him satiated in the story.

Aaron checked into the hotel and asked, "John, before you go, aren't you going to bless my room? Isn't that what pastors do in America?"

"Yes, sir, if that will help you rest well. I will gladly bless your room."

As the men were in the elevator, John thought, *God, how do I pray for this man again? Redemption, please, God. Redemption.*

John walked in the room and put his right hand on Aaron's right shoulder. The men bowed their heads together in prayer, and John said, "Abba, bless this room and bless my new friend with rest." John then said, "I will be back in the morning to have breakfast with you downstairs. Rest well. Good night."

John slid into his car to drive home and shook his head in bewilderment from the day's events. *Why would I have this Jewish man say that prayer? Is he not already God's chosen?*

John joined his wife, Charlotte, at the kitchen table later that evening. There, she reassured him of who he was and that he heard clearly from God on his mission. She also encouraged him to start listening more than he spoke when it came to the Jews and all things Jewish.

Aaron settled into his room and called Ruth. "Hi, my love. I made it to Austin and met John. He is a funny man of God unlike anyone I've ever met. He has brown skin, a medium build, and very short hair. He spoke with an accent and laughed a lot. I asked him to pray twice, once upon arrival and once to bless my room, and I think he tried to convert me to Christianity. I confirm to you that I am still Jewish, but there is something in his tone and confidence that tells me he is a safe, new friend. I trust him and am glad to hear what he has for our people. Good night, Ruth. Thank you for taking care of our children, both the ones you birthed and our city, while I am away."

Next, Aaron called Diana. When Aaron asked for an update, Diana replied, somewhat miffed. "Several constituents came to the office today: a man complaining about noise from a neighboring restaurant, a woman who would like her trash picked up twice per week instead of just once, and grumpy Old Man Cohen who claimed the Goldstein boys were playing soccer in his flower bed again. Nothing of note."

Aaron, tired from his long day of travels, dismissively replied, "That's a great report, and I'm so thankful for your work. Tomorrow,

I will be in one of the most important meetings of my life. Pray that this meeting will result in more blessings for Gibeah!"

The two hung up, and Diana thought, *Why does Aaron get to have all the fun when I am at home dealing with people's silly neighborhood problems?* She thought she took this job alongside Aaron to run campaigns, shape the future for her and her Jewish land, and change the state of the nation for her people. She dismissed the thought, attributing it to being tired, and clocked out early for the afternoon.

The next morning, Aaron met John in the lobby of the hotel where the two had breakfast. For three hours, John read to Aaron numerous scriptures from the Bible that he believed shared the prophetic destiny of Israel—how God would send them back to the land, which had recently been declared through United Nations' Resolutions 181 and 242. "It was God who said I will gather you from the nations and the land belongs to Me. God said the land belongs to Him, and you should not give it away even when you believe it will give you peace with your neighbors."

After poring over scripture and dissecting each passage, Aaron responded favorably to the message. "That makes perfect sense with what I was taught in the synagogue growing up and what I believe as a Caleb that I am called to do in Gibeah. I have been making space to bring the Jews home from Europe and the Middle East, but I never knew God was telling Christians they were part of this, too. I have heard all I need to know. I will return to Israel and let you know when you can come and share this news with the prime minister. Thank you, John!"

The next day, John personally drove Aaron back to the airport for his flight home, and as the men departed, John felt a weight lift off his shoulders. He had delivered the message he needed to deliver.

Within one week of their meeting, John was on a plane to Ben-Gurion Airport in Tel Aviv where Aaron pick him up. With the roles reversed, John chuckled to himself when he got into Aaron's car, thinking he should have Aaron pray for him in Hebrew. He smiled as he rode in the car, taking in the beautiful green and blue sights of the countryside and listening to Hebrew radio broadcasting local news he could not understand.

John was excited to see Israel for the first time and meet with the prime minister in a meeting Aaron arranged. His nerves were jumpy and frazzled, but he maintained a peaceful outer shell to prepare.

Along the drive into the city, Aaron stopped the car abruptly. A military convoy pulled up right on cue to take them on a surprise visit to El Morya. John was frightened by the military vehicles but continued to play cool. On the drive, Aaron said, "Well, John, don't you want to know where we are going?" John replied, "We are going to the prime minister's office." Aaron chuckled and said, "Later, my friend. First, we will visit the very spot God promised Abraham the land of Israel through a covenant. I thought you might need some spiritual inspiration after the long flight, and besides, the prime minister is not available until later this afternoon." Shielding tears with laughter, John felt the weight of being an honored guest in God's homeland as well as God's heart for the people of Israel. He knew how important it was for him to deliver the message he came to share.

After a few minutes of letting John roam and pray at El Morya, Aaron said, "That's enough, John. Get back in the Jeep. I've got something else to show you." The group drove to Shiloh where the Ark of the Covenant rested for nearly 400 years, and John finally let his tears be known.

As he jumped out of the Jeep and put his face to the ground, he heard God say to him, "*Today you shall speak to the prime minister and tell him not to negotiate this land for peace. It is I who brought My people back from the nations, and I've appointed him to protect the land.*"

"John!" Aaron called. "The prime minister's office is on the phone. They are ready for you. Hop in the Jeep. It's time to go."

At the prime minister's office, Aaron and John sat in sturdy chairs across from one of the world's most powerful men. John tapped his foot uncontrollably in nervous anticipation. The prime minister had a kind but stern face. He was a practicing Jew, dedicated to God and his people and known around the world for his fervor. Aaron was an old friend of the prime minister's since they had lived in the municipal project in Tel Aviv at the same time.

John, visibly shaken, shared with the prime minister all he had told Aaron and what he had just received from God at Shiloh.

The prime minister listened with an open heart and responded, "John, have you not read your Bible? Don't you know that we Jews stone our prophets?" The room erupted in laughter, and the prime minister continued, "I'm joking with you, John. Thank you for your determination to share what you feel is important for us. I will make arrangements for you to deliver the message to members of our government and military to get their opinions later this week."

Chapter Eight

After the meeting, Aaron and Diana drove John through the West Bank to Gibeah. Aaron and Ruth enjoyed hosting friends and invited John to stay in their home. They loved showing off the accomplishments and growth of their beloved home. They planned for John to have the best tour, including meeting locals and dining with their friends, to give him plenty of stories to take home.

Ruth welcomed the men and Diana into the Lipkin home with a boisterous "Shalom!" and an oven full of warm pita bread. Ruth invited people into the family well. She was even known for saying, "You are invited to open every cabinet in my kitchen. Anything you can see, you are welcome to."

As they sat down to break bread, Diana faked a headache and excused herself from the table. She felt tension with the Christians and conflict with God. *If we are God's chosen people*, she thought, *then why would He send the Gentiles to share messages from Him? That's not the God I know.* She went home to sleep that evening, feeling angry and distrustful toward the visitors and believing they had ulterior motives for why they wanted to help. She vowed to learn all she could to protect her land and her people, even if she was the only one.

After a lovely dinner of Israeli salad, hummus, and vegetable lasagna, all made from locally sourced ingredients, the couple opened a bottle of local wine for their guest. A robust yet smooth Cabernet

Sauvignon from the Golan, just north of where they were, offered a rich aroma and foundation for John to share the story of Jesus turning water into wine in Cana, found in the second chapter of his namesake book.

John shared that most people focus on the wine itself and miss the depth of the story. He believed it was most important to get to know the winemaker whom Jesus was representing on earth. Jesus did not perform the miracle to simply provide wine for a party but to signify that His Father loves His people and is preparing a wedding feast one day in the future for His children to partake in with the best wine.

Wine-inspired scripture provoked Aaron's heart to remember his youth in the orange grove with his father, Adam. Pruning and continual oversight were key to prevent disease and make the vines the most fruitful.

Aaron shared his love of vineyards and how he planned to inherit them according to Torah prophecy. He shared dreams of his youth that seemed to be revived in his new friend's presence, stories he had not even thought to share with Ruth, buried treasure deep in the soil of his soul. *Unearthed tonight through a Gentile,* he thought.

"Okay, John," said Aaron, "moving on to business." "I think it's time we involve the Christians in funding our next phase of growth for Gibeah."

John replied, "Sir, it would be an honor. I have many friends who are awakening with passion for your nation and your people. Just say the word, and we can begin."

Aaron was elated and agreed to make another visit soon. With a handful of projects on his heart, he prayed to God for funding and waited with gleeful expectation for how He would creatively deliver His answer.

Chapter Nine

Within a month of hosting John, Aaron was on a plane from Tel Aviv to New York. He met with Jewish governmental officials in Brooklyn and New Jersey and was energized by conversations with them and hearing the good work God was doing for them and through them in the United States. They introduced him to Jewish businessmen, some running large corporations on Wall Street with powerful influence in the country despite being only a small percentage of the population. Aaron shared with them that he felt connected because they resembled the people of Gibeah with their tenacity to bring about change.

From New York, Aaron flew directly to Los Angeles where his sister, Tamar, and brother-in-law, Daniel, lived. Tamar had followed her husband to the United States when he was transferred there by the technology firm he worked for. Daniel was a Holocaust survivor and a very strong man. He outwardly exhibited a gentle heart and quiet demeanor. The couple picked up Aaron at the airport and brought him to their home just before sundown on Friday night to settle in for Shabbat. *Tamar made enough food to feed all of the IDF,* thought Aaron, laughing to himself. She and Daniel sat at the table for hours over dinner listening to Aaron share stories of their friends in Gibeah. The couple was amazed at the growth and laughed at what their parents and grandparents would have thought if they were alive to hear the stories today.

Tamar and Daniel were accustomed to making friends with Arab neighbors as they had back home, but they did not have many Christian acquaintances. Tamar shared that recently, Christian women had been reaching out to her in efforts to be neighborhood friends. They seemed genuine and asked good questions about the Torah, synagogues, and Jewish holidays such as Passover. That confirmed to Aaron that God was up to something big with the Christians, which excited him even more for the next leg of his journey, which he had not shared yet with his sister.

Aaron rose early Sunday morning to slip into his freshly ironed Oxford shirt and khaki pants that Tamar bought for him. *Though she is my little sister, she takes good care of me,* he thought as he drove his rental car into the city center. John had arranged for Aaron to speak at a church pastored by his friend Judah, who, along with his congregation, had a special love for Jews, making it a natural first stop on the fundraising tour.

Driving up the Pacific Coast Highway, Aaron rolled down the windows of his rental car to smell the ocean breeze. As he approached the church and pulled into the parking lot, he laughed aloud at the name: The Vineyard Church. *God is teeing me up for success,* thought Aaron. *No one at home would even believe me if I told them.*

Pastor Judah welcomed Aaron into his office before the church service started. They swapped stories about John and all of their love for God. The men had an instant connection, and Aaron knew he trusted John even more in this moment because of the friends he kept. He had a contented and joyous feeling about talking to the people of The Vineyard Church. Judah invited Aaron to sit with his family in the front row of the church. Aaron had never been inside a church but decided to relax into a cozy groove in the pew. After singing a few songs to God, led by a band of 20-somethings with electric guitars and drums, Aaron was a little disoriented but strangely energized.

Judah walked up the five short steps to a stage above the congregation. In his right hand, he held a thick, brown, leather-bound Bible. He placed the Bible on the podium in front of him and delivered an introductory message to the congregation.

"We have a special visitor today, Vineyard. We've been talking about our Jewish brothers and sisters living in Israel, rebuilding the West Bank, or biblical Samaria. I thought it was time that we heard what it is like to be one of the ones told about in the Old Testament who would reinhabit the land for our God. Our friend John introduced me to Aaron Lipkin, mayor of Gibeah. When John told me that the mountain of Gibeah sits near Jacob's well, my heart immediately went to Jesus's encounter with the Samaritan woman at the well. I would like for us all to turn to John 4 in our Bibles and recount the story together, starting at verse seven."

A woman from Samaria came to draw water. Jesus said to her, "Give me a drink." (For his disciples had gone away into the city to buy food.) The Samaritan woman said to him, "How is it that you, a Jew, ask for a drink from me, a woman of Samaria?" (For Jews have no dealings with Samaritans.) Jesus answered her, "If you knew the gift of God, and who it is that is saying to you, 'Give me a drink,' you would have asked him, and he would have given you living water." The woman said to him, "Sir, you have nothing to draw water with, and the well is deep. Where do you get that living water? Are you greater than our father Jacob? He gave us the well and drank from it himself, as did his sons and his livestock." Jesus said to her, "Everyone who drinks of this water will be thirsty again, but whoever drinks of the water that I will give him will never be thirsty again. The water that I will give him will become in him a spring of water welling up to eternal life." The woman said to him, "Sir, give me this water, so that I will not be thirsty or have to come here to draw water."

Jesus said to her, "Go, call your husband, and come here." The woman answered him, "I have no husband." Jesus said to her, "You are right in saying, 'I have no husband'; for you have had five husbands, and the one you now have is not your husband. What you have said is true." The woman said to him, "Sir, I perceive that you are a prophet. Our fathers worshiped on this mountain, but you say that in Jerusalem is the place where people ought to worship." Jesus said to her, "Woman, believe me, the hour is coming when neither on this mountain nor in Jerusalem will you worship the Father. You worship what you do not know; we worship what we know, for salvation is from the Jews. But the hour is coming, and is now here, when the true worshippers will worship the Father in spirit and truth, for the Father is seeking such people to worship him. God is spirit, and those who worship him must worship in spirit and truth." The woman said to him, "I know that Messiah is coming (he who is called Christ). When he comes, he will tell us all things." Jesus said to her, "I who speak to you am he."

Just then his disciples came back. They marveled that he was talking with a woman, but no one said, "What do you seek?" or "Why are you talking with her?" So the woman left her water jar and went away into town and said to the people, "Come, see a man who told me all that I ever did. Can this be the Christ?" They went out of the town and were coming to him.

—John 4:7–30

Concluding the Bible reading, Pastor Judah said, "This is where the first revival occurred, so wouldn't it be like God to bring the Jew and Gentile back together in the very place it all started? This is my theory, as I believe we serve the same God, also named Yahweh, I Am That I Am, Elohim, and the Alpha and the Omega. I will let you all ponder on the beautiful mysteries of our God as I introduce our special guest. Aaron, please join me on stage."

Aaron started his speech with the traditional Jewish Aaronic blessing. "Y'va-reh-ch'cha Adonai v'yeesh-m'reh-cha, Ya-air Adonai pa-nahv ay-leh-cha vee-choo-neh-ka, Yee-sa Adonai pa-nahv ay-leh-cha v'ya-same l'cha Shalom. And now in English, the Lord bless you and keep you. The Lord make His face shine upon you and be gracious to you. The Lord lift up His countenance upon you and give you peace.[1] As Judah mentioned, I am the mayor of Gibeah, a city we built in the hills of Samaria almost 20 years ago. I come representing both the people of Gibeah and the people of Israel. On behalf of them, I thank you for your hospitality in hosting me today. The main goal of our city is to create a space for the displaced and seemingly forgotten Jews around the world so they can come home. These people were cast out of Israel during the Diaspora and now need to return to their biblical homeland. We want to have homes ready to inhabit and vineyards to drink from in celebration. We are making space for our brothers and sisters in the same way our God has made space for all of us in His family."

After sharing his heart for bringing his people home, Aaron proceeded to ask the congregation to fund the news station his team wanted to build. They wanted to take control of sharing their local stories instead of leaving it in the hands of newscasters in Tel Aviv and Jerusalem who did not understand their most provincial lifestyle.

Despite Aaron's broken English, the people of The Vineyard Church were moved to tears at his charismatic delivery. They felt his heart, and it aligned with what God had been putting on theirs. Aaron told them that Gibeah would need $100,000 to start the news station. He invited them to participate in the funding. The audience gave him a standing ovation, and Judah rejoined his new friend on the stage.

Judah put his arm around Aaron and said, "John told us two weeks ago that this request would come from you, and we began to pray and ask God exactly what we should give as a church. The people you see out here prayed individually and gave, as a person, a couple, or a household, for one week. Our leadership asked after praying to return the next week with the money they felt moved to give so we could gather the money in advance for you. What John failed to share with us is the exact amount you would need for the news station. Before collecting from the individuals, our staff met and believed The Vineyard was to give $100,000. We decided if the number that came back from the people was less than that, our general fund would make up the difference.

"When we collected the individual checks from the congregation, they totaled exactly $100,000 and 50 cents, and then you came today and told us that you need that amount to build the station! God provided more than you need." He presented Aaron with a check for $100,000.50 and said, "Here you are, my Jewish brother. God has clearly gone before you and made your path straight.[2] Please take our funds and build your station to share the good news with all who will listen." The people of The Vineyard stood in applause. Aaron cradled his head in his hands and tried to gather his composure. He took the microphone from Judah and said, "You are a blessed people because you support us. God told Abraham very early on in his journey that He would bless Abraham and his people.[3] I pray the Lord blesses each of you today!"

Aaron made several other stops along the California coast and in Texas where John had opened doors. He stayed with local families and made many new friends along the tour.

The night before returning home, Aaron called Ruth to say, "Darling, you're not going to believe this, but the Christians are more generous than we ever imagined. I am coming home with half a million dollars earmarked for the news station, hospital, and recreation center. I am one happy man."

Ruth laughed on the other end of the line, embracing one of their children while looking at a pot of stew boiling on the stove. "Aaron, I am so proud of you. It is an honor to be your wife and pioneer this city. The kids and animals have been a little trying this week, but it's nothing I cannot handle. Diana just left the house, and we talked all about the plans for the upcoming reelection campaign. We all miss you and are excited to have you home!"

Chapter Ten

Benjamin was a young conservative Jewish man from New Jersey with a beautiful, young Jewish wife. After making Aliyah, the absorption birthright program, Benjamin was assigned to work in the mayor's office in Gibeah. Benjamin had ruddy reddish-brown hair, brown eyes, and glasses that gave him an inquisitive look. He was slender and towered above Aaron, standing more than 6 feet tall. Aaron mentored Benjamin, and Benjamin equipped Aaron with creative ideas and advocated for him both in Israel and Gibeah. Benjamin was fearless when it came to talking to politicians, donors, and constituents. Aaron quickly saw Benjamin's potential for greatness and wanted him to stay permanently to help build the city.

One morning over coffee, Aaron presented Benjamin with a verbal offer to be his right-hand man and personally help him build the city going forward. Benjamin agreed to the arrangement and dedicated his life and family to the Gibeah cause. Benjamin joined just before Aaron's reelection vote, which he won by a landslide.

Aaron and his team went to the ribbon cutting for the news station. He asked Diana to contact all media outlets in Israel to inform them that the broadcast station was live. Soon, Gibeah was broadcasting its weather and news highlights to citizens throughout Israel.

Gibeah was well staffed, well spaced, and now well equipped to bring the Jews home. Aaron reveled in these thoughts as he sat on his desk chair. He let his mind wander back to the stories his father told him of the empty passenger jetliner that flew from Tel Aviv to Yemen over enemy territory in the 1950s. The Yemenite Jews were displaced from the lineage of King Solomon and kept the exact language and tradition of his temple recorded in the Bible.[1]

In modern day, they found themselves enslaved by the Arabs in Yemen and unable to leave. The problem was that these Yemenite Jews had never seen an airplane before. They lived in a society without television or radio. They were starved for education, communication, and technology. When a plane finally arrived carrying Jewish government officials to fly them home, they responded, "No, thank you. Please stop trying to kill us like the rest of the world."

They were terrified to get on the airplane and thought it was another ploy to exterminate them. The Jewish government brainstormed ideas to make flying comfortable and remembered the Arabian Nights story. After replacing airplane seats with plush rugs, Operation Magic Carpet was a documented success. Aaron remembered this story while planning the new Yemenite synagogue. He was grateful that Gibeah would host the presence of God as they did in Solomon's day through Yemenite tradition.

The next group of Jews Aaron hoped to bring home were the recently liberated Russian Jews. Once they were allowed to leave the country, the Israeli government had planes waiting to gather these men and women and bring them to Israel. Most of these people were well educated and had strong science backgrounds. Aaron wanted to host these people in Gibeah. He enlisted a group of men to drive to the Ben-Gurion Airport and meet the Russian Jews at the baggage claim. Many arrived in Israel with only a suitcase of clothes and important documents they had saved. These people came to find a better life, and they found it in Gibeah. Aaron and David lined up jobs for them and homes that fit the desires of these newcomers. The Russians were grateful for the homes and jobs. They integrated well with the other Jews who had been previously scattered. They felt it was a homecoming and gladly helped Aaron grow the city to make space for even more dispersed and forgotten Jews. Adding these international transplants doubled Gibeah's population.

Though only Jews could live in the city of Gibeah due to original zoning, Aaron wanted to create a way for them to work and interact alongside their Arab and Christian neighbors and visitors.

Gibeah University became the next item on the agenda as privately funded universities were not regulated by the government. This would be a place where all could come and integrate. What he did not realize at the time was that inclusive behavior built a natural shield that protected the city as geopolitical pressures heightened. All were welcome, which pleased the Palestinians, the Israelis, and the international community.

Chapter Eleven

John returned to Israel as often as possible to visit his friends and make new ones. On one of these trips, Aaron invited him to speak at Gibeah City Hall to share with government officials that Christians in America desired to send more money and start making trips to visit Gibeah and see what their dollars had funded. As he concluded his speech, he spoke words foreign to him. "Because you have opened the doors to the Christians, God is going to plant a Smart Vineyard in Gibeah."

His eyes grew large as the words confidently departed his lips, and he thought, *Thank goodness I'm jumping in a car for the airport now and will not have to discuss with anyone what this actually means. What in the world did that mean, God?*

John said a swift good-bye to Aaron and went to the airport. At the airport in Tel Aviv, John boarded a flight scheduled to land in Austin after several layovers, giving him plenty of time to ponder his words. On the first layover in London, he disembarked and walked through the crowded terminals of Heathrow, unable to shake the phrase *smart vineyard* from his mind. Just as he said the words under his breath, he passed a newsstand with a global economic magazine staring him in the face. Its cover read *Biodynamic Farming, It's Here.*

Well that sounds like a pretty smart idea for a vineyard to me, he thought as he purchased the magazine and glanced through its pages.

As John boarded his next flight, he learned all that he could about biodynamic farming. Airborne, John consulted with God about His plans for Gibeah and how he could play a part in its further development.

Right on time, the day after John arrived in Austin, Aaron called him and said, "My dear friend John. I trust you made it home. Thank you again for coming to visit. Now what in the world is a smart vineyard?"

John held the phone and laughed into the receiver for an uncomfortable period of time. When he composed himself, he replied, "I found this magazine that I believe holds a key to unlocking the vineyard. I learned that there are vintners in other parts of the world who are pioneering grape growing through sustainable and eco-friendly means. I believe Gibeah is to follow this model and replant the vines of your ancestors. I've already put in a call to a consultant who has started on a feasibility study. He said he would like to meet you and take soil samples."

Aaron said, "No, John. God told *you* to turn Gibeah into a smart vineyard! It's your assignment, and *you* are to organize and pay for it."

"But I don't have the money to pay for it. I thought I was just the messenger," John said.

"John, you've always told us that where there is a vision, there is provision. If you know the God of Abraham, Isaac, and Jacob gave you this message, then what is time and money?"

"You *have* been listening to my sermons!" John said. "You are right, my friend. I will believe that God will take care of it. I will do my part, but I need all of you to do your part over there and pray."

"No, John," Aaron replied. "You pray and you bring us the money and the plan."

"God said he would put His throne on the Temple Mount, which means it is a local call for you. It's long distance for me. So you make the local call."

The two men bantered about who, in fact, was actually closer to God and landed on John owning the project.

John hung up the phone and walked into his kitchen to find Charlotte and share the story. The two laughed about God's vastness and their relationship with Israel over a bowl of hummus she made from fresh tahini John bought at the Mahane Yehuda market in Jerusalem just days before. Bite after bite as they scooped up the hummus with warm pita bread, the couple sat deeply in silent thought for a while, pondering the size of the request.

Charlotte broke the silence. "Remember all that God has done for us and how much He loves Gibeah? He will bring the money and the plan. You have nothing to worry about."

Chapter Twelve

Aaron visited the United States as often as he could, building relationships with American Jews and Christian organizations. Christian groups were catching wind of their Jewish brethren in need of funding and support, and some early adopters were glad to help. There were networks such as Good News Network and Trinity News Talk that picked up the cause and solicited audiences for funding to send abroad. Once the Jews made the mainstream Christian news broadcast, the floodgates opened, and support rained down on individual churches wanting to partner with organizations in Gibeah.

It was as if a veil had been lifted, and they could see clearly that the Jews did not, in fact, "miss it" as they had previously been taught. These Christian groups replaced old ways of thinking with new perspectives that resulted in blessing, honored the Jews for writing the Bible, carried out the traditional feasts and festivals appointed by laws that are recorded in the book of Leviticus, and returned to the Promised Land. They believed this would usher in the coming of the Messiah—the first time for the Jews and the second time for the Christians. Bridges between the two groups were being built. Aaron rode the wave of momentum and spoke to anyone in the United States who would listen.

Gibeah received millions of dollars in funding from Christians in America. The Christians were looking for a safe and trustworthy place to send their money. The group Americans for Gibeah was formed with some of its largest supporters hailing from New York, Texas, and California. Aaron entrusted the collection and maintenance of this organization to a group of friends in California. He thought, *This way, long after I am gone, I will ensure the bridges between the Christians and Jews will stay intact.*

With unprecedented reserves, Gibeah experienced accelerated growth. Aaron spent his time between trips to the United States overseeing projects in Gibeah and preparing for the next mayoral election. That required him to lean on his people like never before, trusting them with more responsibility and even greater professionalism.

Diana singlehandedly led the upcoming campaign, Benjamin was given free rein to coordinate Gibeah projects, and the Americans for Gibeah increased its marketing efforts and managed Aaron's tour calendar. Ruth continued growing and managing the zoo, both at the local establishment with animals and at their home.

To maintain the delegation of global tasks, Aaron kept multiple cell phones on his belt at all times. He could be reached from anywhere in the world, including his home. The stakes were raised and stress was heightened in the Lipkin household and the mayor's office. Aaron noticed a negative change in his diet, sleep, and relationships but thought, *If I can just get through this election, things will calm down and go back to the way they were. I will take Ruth and the children on a vacation to a European beach when they are out of school for the summer. We will meet David and his family. It will be just like old times.* With this thought, Aaron drifted off to sleep next to his beloved Ruth.

About an hour later, in the middle of the night, Aaron received an unexpected phone call that jolted him out of bed. He grabbed the ringing phone out of the pile of many and ran downstairs to answer. As he stepped outside onto the front patio so he would not disturb Ruth, an unfamiliar man spoke.

"Hello, Aaron. You must stop the building of Gibeah immediately and surrender the land to the Palestinian authority. It does not belong to the Jews. If you do not obey, we will find you. We will kill your family and launch your city into flames. We will ensure the land will be restored ourselves."

Aaron yelled, "You coward! Show your face to me! Do not make empty threats through the phone! We will never stop building Gibeah! This is our land!"

Aaron heard a click on the other end. He walked back into the house for a cigar and brewed a hot cup of coffee, knowing sleep would be impossible for the rest of the night. He grabbed the secure cell phone and called Israel's Minister of Defense.

"I just received a death threat from a man speaking Arabic. I know you've allocated us IDF soldiers, but I need you to double their presence starting at sunrise. We are growing quickly and need protection."

"I hear you, Aaron, and I know we've been friends for a long time, but the allocation and ownership of land in the West Bank is a very tricky subject. We are working with the United Nations and the United States government to navigate the territory. We are also talking with our neighboring countries that are increasing pressure for a more equal Jewish and Palestinian state. I am still a Zionist and in the trenches fighting with you, but I have more pressure than we experienced in the 1970s when we first started."

"I know, dear friend," Aaron replied. "I know you are under a lot of international pressure, but we must never give up the land. The United Nations and the Americans have no business interfering with our boundaries. The Arabs are our cousins, and we don't need strangers from other cultures telling us how to run our home. Now there are creative ways to partner with the Arabs and the Christians. I've done it in the orange groves, and I've done it in America. I can do it in the West Bank. We just have to keep the land. Please give us more soldiers to better protect us. This man threatened my home and the people I love. Aaron Lipkin does not back down. We knew there would be giants in this land. Our forefathers encountered them and crushed them. Bring me the giant killers."

"I will see what I can do, Aaron. Try to get some sleep."

Aaron took a stroll through his neighborhood, picking up stray trash on the ground. As he paced, he remembered his mission.

"We can do this, God" said Aaron aloud, "This is Your fight and Your land. I am your servant, and You will protect us, Yahweh. Blessed are You, O Lord."

After talking with God, Aaron went into the house and e-mailed John and Judah, asking them to pray for resources and safety.

When Ruth awoke, Aaron shared the disturbing phone call with her, just as he had shared all other details with his beloved. He trusted and valued her opinions on all matters.

"Aaron, thank you for fighting for us," Ruth replied. "I love you and will stand strong with you as we grow Gibeah together."

Aaron let himself nestle gently into her arms and cry. After a few minutes of embrace, he composed himself, got dressed, and kissed Ruth good-bye.

"Have a good day, my love," he said, and then went to check in with Diana for his daily mayoral itinerary.

This election was different than the others with more media exposure than ever before. "In the early days of Gibeah," Aaron said, reminiscing over a cup of tea with Diana at the office, "we handed out flyers and went door to door, connecting with the people. I feel like now I am on every television channel I turn on, both local and global. I feel exposed and polarizing, which means we are onto something, Diana. People either love or hate us. We are going to win this one like all the others and keep building our city."

When Diana did not respond, Aaron asked, "Diana, is everything okay? I just realized I have not checked in with you in a while to see how you are handling the increased exposure. How are you doing?"

Diana seemed distant as she shuffled papers quickly. "I'm fine, Aaron. I'll be fine. I'm just ready for this one to be over."

Aaron looked at her quizzically and made a mental note to buy her a spa package from an exclusive spa in Tel Aviv and give her extra time off from work when this election was over. He wanted to make sure his staff and his family knew how much he loved them and appreciated their efforts for their joint cause. He thought, *Diana knows how much I value her. She is a great assistant, and I do not know what I would do without her.* He thanked God for bringing her to his team, and he went about his day, fulfilling the items on his agenda.

Diana checked Aaron's e-mails and found an inbound prayer from John and Judah. Both men shared their hearts for Gibeah and asked God to protect its development and the people who lived there. Diana deleted the e-mails in haste and spit at her computer screen.

Under her breath, she said, "We do not need your help. This is our country, our city, our cause. Leave us alone. You do not belong here in what we've built."

Aaron won the pending election, but it was the closest vote count since the city's inception. Aaron and Ruth again watched the news report the results, surrounded by friends and family as was now the tradition. As everyone congratulated Aaron and made their way from the restaurant to their homes, Aaron looked around and said, "Ruth, have you seen Diana? I have not seen her all night."

"No, Aaron. Now that you mention it, I have not seen her. I hope she is all right."

Aaron went home to sleep but could not sleep with thoughts of Diana on his mind. She had been acting strange in their office and now skipped the party that celebrated their victory. Something was very wrong, and he needed to know that she was okay. *Had she been abducted by the man that threatened him on the phone? Was she sick? Was she in trouble?* He got out of bed and wrote her an e-mail on his desktop computer.

"Diana, please meet me for breakfast in the morning at Jericho Café. We missed you tonight, and I want to make sure you are all right."

The next morning, Aaron sat in his regular booth at Jericho Café and ordered his favorite bagel and lox, something he had grown to love in the United States. As an hour passed and Diana had yet to arrive, Aaron began to fear that something terrible had happened to her. As he picked up the check to leave, his mayoral election opponent walked into the restaurant and sat across the booth from him.

"Aaron, congratulations on a close race. I was sure I would have won this one with all the extra intel I had."

Aaron looked perplexed and replied, "Intel?"

"Yes, Aaron, intel. Oh, you haven't heard. Diana switched sides a long time ago and joined my team. She's been stealing information from your files and e-mails. She also had your phones tapped and shared everything with my team in hopes to defeat you."

"But I don't understand," replied Aaron. "She has been with me for many years. She shares the same heart for our beloved Gibeah. She is a part of our family."

His opponent informed Aaron that Diana did not like all the attention Aaron was giving to the Christians and did not believe they belonged in the city. He slyly alluded to bribing her for information and how she chose to sell Aaron out for only a few thousand shekels.

Aaron left the restaurant and went straight home. He opened the bedroom door and collapsed into Ruth's arms yet again. She held him tightly as Aaron explained Diana's betrayal. The Lipkins never saw Diana again, and her name and memory were erased from the Lipkin household. Aaron took his first day off from work since he was a young man, simply to rest. To his recollection, he had never felt the pain of betrayal so deeply before.

Chapter Thirteen

Back in the office, Benjamin was doing the work of a full staff, and Aaron was back on an airplane to the United States to celebrate the election win with his friends while networking with new church partners. After spending time in California and Texas, Aaron felt unusually exhausted on the last night of his trip. At dinner that night with one of his favorite rabbis in Manhattan, Aaron broke out into a sweat, immediately soaking his clothes. Ending the dinner early, Rabbi Isaacson would not let him return to his hotel room alone and decided to have a taxi take them to his home in Brooklyn. In the taxi, finishing their conversation, Aaron stopped talking mid-sentence. Rabbi Isaacson looked over to find Aaron collapsed on the car window. He yelled, "Driver, take us to the emergency room!"

After Aaron was in the hospital for a day on fluids and medication to stabilize his heart, his doctor suggested that Ruth fly to New York while they ran tests to find the reason for the collapse. Aaron boldly told the doctor that he was fine and asked for his phones.

"Doctor, I am like Caleb of the Bible. I am fearless. Now release me from this hospital. I need to get home and build my city."

The doctor told Aaron to relax and pumped his IV with a strong sedative to help him sleep.

Ruth dropped the children off at David's house and boarded a plane to New York. When she arrived at the hospital after 24 hours of travel, Dr. Patel, cardiac specialist, reassured the Lipkins that fatigue was a common problem for politicians and that he would be running a complete blood count, chemistry panel, and urinalysis, checking Aaron's blood glucose levels, and doing hormone testing to diagnose him properly. Aaron slept for two days while the couple waited for the test results.

Ruth called John and Judah to ask for prayers. She knew this is what Aaron would have done—sought help from the Christians—and since he could not do it at that moment, she would do it on his behalf. Exhausted, Ruth laid down on a cot next to Aaron, covering her body with a warm, heavy blanket. She grabbed the Torah lying next to his bed and read from somewhere in the middle until her eyes glazed over and she joined him in a deep sleep.

The couple awoke the next morning to a somber-looking Dr. Patel who was holding a manila folder. He said, "Mr. and Mrs. Lipkin, I have the results. It appears that Aaron's fatigue was caused by stress cardiomyopathy, or better known as broken heart disease. The panels showed a blood clot in his heart. We recommend a strict change in diet and exercise immediately and that he take a sabbatical from government work for a time to heal."

Ruth responded, "But I don't understand, doctor. My Aaron has always been in such great health! It is not possible for him to have a heart malfunction."

"I do not know the specifics of your husband's case, but I've seen this before in some of the strongest and healthiest men. Sometimes a broken heart can crush the soul, letting wounds fester and allowing disease to come in."

Diana, thought Ruth. *This is all her fault.*

Aaron and Ruth returned home to begin heart treatments in Tel Aviv. Once at home in Israel, they were besieged with cards and balloons from friends around the globe. Some family and local visitors were allowed to see Aaron, but only if they had good news to share that kept him smiling. Everyone marveled at Aaron's endurance and zeal for life. He told all who visited that he would not be defeated by a silly upset. He would live and not die in order to recount the deeds of his God.[1]

Aaron was overwhelmed with joy when his friends John and Judah trekked across the world to visit their very ill friend. After chatting with him during a treatment, they asked Ruth if they could drive him home and spend some time with him that evening. Ruth was grateful for their volunteer efforts and gladly obliged. She cooked the men a meal and brewed fresh decaffeinated tea before retiring with the children early that evening. John and Judah ate dinner and then sat with their friend in his living room. Aaron loved his wife's cooking but could barely stomach the smell of the warm food wafting through the house that evening.

Though Aaron was covered in a cozy blanket and in a restful position, he never exchanged his pressed Oxford shirt and khaki pants for pressed fleece pajamas until he was ready for bed. His appearance was very important to him, and he believed it helped him fight. He was always ready to work in a moment's notice if his constituents or his God needed him.

John said, "Aaron, when I first met you in person, you asked me to pray for you. You thought that was what preachers do. You were right. It is what we do, and Judah and I would like to ask our God to heal your heart from betrayal. We believe it is not His will for this health problem to be in your body, and we believe you can be healed from it. Would it be all right if we pray for that?"

"John, I trust you, my dear friend. You ask our God for whatever you think I need."

John and Judah knelt down beside their friend and held his cold and clammy hands. They asked God to heal his heart and his entire body. They asked God to cover him in peace that surpasses all understanding. After they prayed, they assisted their friend to his second-floor bedroom and laid him in bed next to Ruth.

With that, John and Judah called for a taxi and made their way to Ben-Gurion Airport that evening to fly home to the United States.

Chapter Fourteen

After six months of sabbatical rest, daily physical therapy, and reluctantly giving up all but one of his cell phones, Aaron had fully gained back his strength. He wanted to be healthy enough to attend the grand opening of the Gibeah Performing Arts Center. This project was most dear to his heart out of all his building projects. He always supported the arts and believed their beauty helped people focus on love over hate. The Tel Aviv doctors allowed him to attend if he promised not to overexert himself or stay up too late.

As Aaron walked through the crowd, blowing kisses to the citizens of Gibeah on this monumental opening night, he received praises and blessings from all in attendance. It was a sweet reward for his years of lobbying to have this art house constructed. Reporters from the Gibeah news station asked him how he was able to push through the pain for an event like this. Aaron spoke into the camera, "Being active is my rest. If I'm not active, I feel that I'm not really alive."

The event was well attended and covered by media personnel from around the country. No one believed there would be arts again in the hills of Samaria, and all had to see it for themselves. The performances that evening were a conglomerate of talents. From traditional Israeli dancers to a stand-up comedian, many constituents were seen, heard, and known by the public for the first time.

When the performances were over, Aaron slowly climbed the seven steps to reach center stage. Before his speech, he raised his glass of locally sourced Merlot and pronounced, "Cheers! L'Chaim! Blessed are You, Lord our God, King of the universe, who has kept us alive, sustained us, and enabled us to reach this season, the Gibeah Performing Arts Center inauguration. Amen!"

The people followed their leader, raised their glasses, and cheered loudly. Aaron breathed deeply and took in all the sights and sounds of the atmosphere. He felt satisfied by the accomplishment of making this dream a reality. The audience cheered even louder when he called the first lady of Gibeah to the stage. The couple held hands and took their bows.

Aaron grabbed Ruth, hugging her, and said, "It's a dream come true! We did it!"

As he let the tears stream down his face in public, Ruth hugged him tenderly and beamed with joy, forgetting about the pain of the past for a few glorious moments.

"Aaron, you've done it. You've brought the Jews home and given them a seat at your table."

Aaron felt a new knowledge of home. A deep sense of belonging arose in his soul. He only wished his own father, Adam, could have been there to see it all.

The Gibeah Performing Arts Center was built next to the hospital that was now fully operational. Aaron's dream to create a place for babies to be born, for day surgeries to be performed, and for general accessible care needs was realized. David was on the board at the hospital and led with excellence and valor.

The hospital was across the street from the fire station. Aaron remembered its beginning well. He shared with a Texas friend that the Israeli government was not able to provide ambulances to anyone in the West Bank due to international pressure prohibiting growth in the region.

"Not on my watch, Aaron," he replied. "I will order three American ambulances to be shipped to Gibeah within the month that can help you start a fire station and mobile ICU."

Aaron was grateful for all these places, knowing they would decrease the mortality rate of Gibeah, which was still low compared to the rest of the State of Israel.

Also nearby were the grocery store, optometry clinic, municipal library, recreation center with a swimming pool, tennis courts, and a Holocaust museum.

One of Aaron's favorite projects was the IDF resting corner where daily meals were provided to soldiers patrolling and visiting the city. The men and women serving would receive food, water, shelter, and prayer, if desired. Aaron visited the IDF soldiers often, reminding them what a positive impact they were making on the country.

Another project accomplished during Aaron's time as mayor was the establishment of Gibeah University. He wanted to create a space for young people to be educated as he had been in Tel Aviv. Aaron envisioned this part of the city as a place where Jews and Gentiles alike could come together and study whatever they desired. It was a place for them to unite and dream of future inventions, technology, and cities that would change the world long after he was gone. With 15,000 students, it was now one of Israel's largest public universities. People were reverse-commuting from the major cities to attend.

In order for Gibeah to be self-sustaining, the city would need to bring the trades that were practiced in Tel Aviv and Jerusalem and expand commerce. Aaron's first vision of a business center included a few Jewish businesses, but his vision expanded one afternoon on a bus ride home from Tel Aviv. Striking up a conversation with an Arab traveler, Aaron learned that the man commuted daily from the West Bank to Tel Aviv where he worked at a semiconductor plant. The thought struck him, *Why does the business center need to be exclusive to Jewish businesses and businesspeople? What if a section of the park was dedicated to Arab businesses and businesspeople, and the two could live in peace working alongside each other?* Today, the business center has Jewish businesses within Gibeah's city limits and Arab businesses on the other side of the city border. In partnership, the two groups share utilities and labor, as well as manufacturing equipment for the State of Israel.

One concern Aaron had was for the water filtration system. He feared the Arabs might try to poison the water supply as they had done in other communities in the past. Aaron knew the industrial park would unify the two groups, but he asked relocated Russian scientists to create a water filtration system that would feed the Arab communities but allow Gibeah to control the source. The result was a hydroelectric science park that rerouted the water to flow uphill from the city of Gibeah to surrounding locations that supplied Arab residences and businesses. If the Arabs tried to poison it, they would only be poisoning themselves as the water moved upstream toward their people.

Aaron accomplished these projects with the support of the citizens of Gibeah and his American friends. He was grateful for Christian donations but desired for Gibeah to be self-sustaining soon. Though he still had many friends inside the Israeli government, they were more focused on the increasing tensions caused by an increasingly global geopolitical world. Aaron was afraid the United Nations was negatively influencing both the Israeli government and the governments of all member nations to control their small country. He could not ever comprehend, however, why the rest of the world was so concerned with a country that had the same size land mass as New Jersey. Jews comprised less than 2 percent of the world's population. *Why did the world care what happened to them?* Aaron wondered.

After recovering from his heart blockage, Aaron spent two years growing Gibeah at a rate he had only dreamed of at its prior peak. It became his first love and his only thought. He knew his job was to make sure the people of Gibeah had everything the people of Tel Aviv had.

Chapter Fifteen

Aaron placed a call to John's office requesting him to visit, to which John replied, "Aaron, I plan to be in Gibeah in a few months. What is the rush?"

Aaron said to his longtime friend, "John, I have a surprise for you. Call Judah, and you guys get here as soon as you possibly can."

The two men landed in Tel Aviv just two weeks after the call. Aaron picked them up at the airport and drove them to Gibeah as he had many times before. "What's this surprise you have for us, Aaron? Why the rushed trek to the Holy Land?" inquired Judah.

John said from the back seat, "Yeah, what's the rush, Aaron? The last time you gave me a surprise, I ended up in a military convoy on my way to a contested religious site. I wore my running shoes today just in case I encountered trouble."

"After all these years, you guys don't trust me?" retorted Aaron. It is a great surprise. You will rest at the house tonight with Ruth and me, and tomorrow we will go on an adventure into the hills together. That is all I am sharing.

The next morning, John and Judah woke earlier than the rest and hiked to the tallest point of the Samaritan hills. There they took in the sights, sounds, and smells of the bustling small city and watched the commuters leave for work and the children walk to school. They prayed for the people of Gibeah, believing they were the first of the Jews to come home to Jacob's well.

John said, "Judah, this morning I feel it's important to read from the book of Romans, the passage in the 11th chapter that, I believe, tells us more of why we are here. It reads like this:

> *I ask, then, has God rejected his people? By no means! For I myself am an Israelite, a descendant of Abraham, a member of the tribe of Benjamin. God has not rejected his people whom he foreknew. Do you not know what the Scripture says of Elijah, how he appeals to God against Israel? "Lord, they have killed your prophets, they have demolished your altars, and I alone am left, and they seek my life." But what is God's reply to him? "I have kept for myself seven thousand men who have not bowed the knee to Baal." So too at the present time there is a remnant, chosen by grace. But if it is by grace, it is no longer on the basis of works; otherwise grace would no longer be grace.*
>
> *What then? Israel failed to obtain what it was seeking. The elect obtained it, but the rest were hardened, as it is written,*
>
> > *"God gave them a spirit of stupor,*
> > *eyes that would not see*
> > *and ears that would not hear,*
> > *down to this very day."*
>
> *And David says,*
>
> > *"Let their table become a snare and a trap,*
> > *a stumbling block and a retribution for them;*
> > *let their eyes be darkened so that they cannot see,*
> > *and bend their backs forever."*

So I ask, did they stumble in order that they might fall? By no means! Rather, through their trespass salvation has come to the Gentiles, so as to make Israel jealous. Now if their trespass means riches for the world, and if their failure means riches for the Gentiles, how much more will their full inclusion mean!

Now I am speaking to you Gentiles. Inasmuch then as I am an apostle to the Gentiles, I magnify my ministry in order somehow to make my fellow Jews jealous, and thus save some of them. For if their rejection means the reconciliation of the world, what will their acceptance mean but life from the dead? If the dough offered as firstfruits is holy, so is the whole lump, and if the root is holy, so are the branches.

But if some of the branches were broken off, and you, although a wild olive shoot, were grafted in among the others and now share in the nourishing root of the olive tree, do not be arrogant toward the branches. If you are, remember it is not you who support the root, but the root that supports you. Then you will say, "Branches were broken off so that I might be grafted in." That is true. They were broken off because of their unbelief, but you stand fast through faith. So do not become proud, but fear. For if God did not spare the natural branches, neither will he spare you. Note then the kindness and the severity of God: severity toward those who have fallen, but God's kindness to you, provided you continue in his kindness. Otherwise you too will be cut off. And even they, if they do not continue in their unbelief, will be grafted in, for God has the power to graft them in again. For if you were cut from what is by nature a wild olive tree, and grafted, contrary to nature, into a cultivated olive tree, how much more will these, the natural branches, be grafted back into their own olive tree.

Lest you be wise in your own sight, I do not want you to be unaware of this mystery, brothers: a partial hardening has come upon Israel, until the fullness of the Gentiles has come in. And in this way all Israel will be saved, as it is written,

> *"The Deliverer will come from Zion,*
> *he will banish ungodliness from Jacob";*
> *"and this will be my covenant with them*
> *when I take away their sins."*

As regards the gospel, they are enemies for your sake. But as regards election, they are beloved for the sake of their forefathers. For the gifts and the calling of God are irrevocable. For just as you were at one time disobedient to God but now have received mercy because of their disobedience, so they too have now been disobedient in order that by the mercy shown to you they also may now receive mercy. For God has consigned all to disobedience, that he may have mercy on all.

Oh, the depth of the riches and wisdom and knowledge of God! How unsearchable are his judgments and how inscrutable his ways!

> *"For who has known the mind of the Lord,*
> *or who has been his counselor?"*
> *"Or who has given a gift to him*
> *that he might be repaid?"*

For from him and through him and to him are all things. To him be glory forever. Amen.

—Rom. 11:1–36

John said, "We have a lot to learn from this scripture. We do not need to focus our energy on evangelizing the Jews. We need to focus our energy on loving them. We were grafted in by the grace of God.

It's our love and commitment to unite Jew and Gentile. Then they will see the fullness of our Beloved."

Hearing the sound of hands clapping behind him, John turned around to see Aaron walking toward him.

"Bravo, John! That was a great speech. I agree. You guys do need to love us more."

With a hearty laugh, the men each put a hand on the other's shoulder and looked at the city below. Aaron said, "Finish your Bible study quickly. It's time to show you my prized project. Also, I'm glad you are wearing comfortable shoes."

The men winded down the hill in Aaron's car. John and Judah knew well enough by this time not to ask questions of where they were going or what they would do when they got there. When they arrived at their destination at the base of the hills, the Christians found several acres of tilled soil, a few shovels, and some greenery draped across the dirt.

Aaron stepped out of the car and pointed at the shovels. "The timing of your speech on top of the mountain about grafting could not have been more precise. It is as if you knew what you were going to do today. Christians, today you will help Gibeah break ground on the vineyard. My father told me when I was a teenager that I would be the one to replant vineyards in the hills of Samaria, the vineyards that had been lost in the war with our enemy. I've been wanting to break ground on the vineyard and prayed about who to invite to join me. When I met you, John, I heard God say that you were the man. A tree or vine cannot produce the same true fruit unless a new branch is grafted into the plant's stock to nourish and support it. You gentlemen are representative of that branch, and today you will connect Jew and Gentile, fulfilling a biblical mystery these hills have held for thousands of years."

With that, the men picked up their shovels and broke ground on the vineyard.

Chapter Sixteen

Seven years after his Manhattan diagnosis, Aaron took a morning stroll through the vineyard, pulling an occasional weed and checking on the drip irrigation system. Though feeling peaceful in the vineyard, he had taken on much stress again preparing for another upcoming election.

His thoughts shifted toward Diana as they often did in election season, and Aaron felt his blood pressure rise.

Under the beautiful sunny day with few clouds and a slight breeze, Aaron's heart gave out. In a fight for his life, Aaron grasped at the vines for help, but the damage to his wounded heart was already too great.

Chapter Seventeen

Shortly after Aaron Lipkin's death, his dearest friends—both Israeli and international—honored him at a memorial service. Members of the Knesset and high-ranking IDF officers were in attendance. The event was covered by national television stations and mentioned in global newspapers.

Ruth, John, Judah, David, Benjamin, and the Lipkin children all sat in the front row and wept gently as the eulogy was given by the Yemenite synagogue rabbi. While looking at each other, they realized what a beautiful family Aaron had built through kindness and trust.

John, who was asked to share stories of his friend, said, "This man, Aaron Lipkin, was the most unlikely hero for a friend. I never dreamed I would know Jews, let alone be best friends with one. He was always full of surprises and gave life his most valiant effort. I trust as he is in heaven today, looking down on all of us gathered here, he is crafting plans with God for Gibeah's next project. But the most remarkable part of Aaron's legacy will always be the Smart Vineyard. With winemaking now in full production on a sustainable and eco-friendly farm, the world can taste and see that the Lord is good.[1] Aaron will always be with us. We will drink this wine that brought Jew and Gentile together, this wine that fulfilled biblical prophecy of old, this wine that celebrates the story of our God. With our hearts full of memories, L'Chaim!"

Chapter Eighteen

On the first anniversary of Aaron's death, Ruth trekked to the gravesite to remember her love. Wearing a lovely flowing linen dress, she approached his grave with as much devotion as she had on their wedding day.

"Hello, my sweet Aaron," she said as she got close.

With eyes clouded by tears, she looked down to see the all-too-familiar green grass and gray stone epitaph marking his life with Hebraic adoration. However, this time, she was also greeted with a single purple flower growing next to the headstone where Aaron's body lay. *This was impossible,* she thought. No one has planted flowers here. Ruth wiped the tears from her eyes and looked again. The flower stood at attention as it shot through the manicured grass and pointed toward heaven, illuminated in the afternoon Israeli sun. It was a beautiful sign of life on what was once believed to be the Mountain of Death.

Ruth realized that moment that this flower was like the cactus in Eliat, and that cactus was like the orange grove in Moshava, and the orange grove was like their beloved city of Gibeah. Aaron had a special touch and knew how to make dead things come to life. Where everyone else saw death, Aaron saw the opportunity for new life. It was his mandate from heaven. Ruth knew her Aaron had gone to be with the creator who had given him the assignment.

Numbers 14:24 reads, "But my servant Caleb, because he has a different spirit and has followed me fully, I will bring into the land into which he went, and his descendants shall possess it."

Ruth lifted her head to heaven and smiled. She knew her Aaron was shining down on her with wild and free long hair and bronzed skin. *Though his body was dead,* she thought, *he had never been so full of life. He would continue to make things grow even from the grave.* This thought gave her new hope. Though she missed Aaron so deeply that the core of her being still ached, she was once again excited for the growth of their city, the city of Gibeah that they and all of Israel had built together.

Notes

Setting the Stage

1. "Balfour Declaration 1917," *Yale Law School: The Avalon Project*, http://avalon.law.yale.edu/20th_century/balfour.asp.

2. UN Security Council, Resolution 242, S/RES/242 (November 22, 1967), https://unispal.un.org/DPA/DPR/unispal.nsf/0/7D3 5E1F729DF491C85256EE700686136.

Chapter Three

1. *For in seven days I will send rain on the earth forty days and forty nights, and every living thing that I have made I will blot out from the face of the ground* (Gen. 7:4).

2. *At the end of forty days they returned from spying out the land* (Num. 13:25).

3. *And the Lord's anger was kindled against Israel, and he made them wander in the wilderness forty years, until all the generation that had done evil in the sight of the Lord was gone* (Num. 32:13).

4. *When he was forty years old, it came into his heart to visit his brothers, the children of Israel. And seeing one of them being wronged, he defended the oppressed man and avenged him by striking down the Egyptian. He supposed that his brothers would understand that God was giving them salvation by his hand, but they did not understand. And on the following day he appeared to them as they were quarreling and tried to reconcile them, saying, 'Men, you are brothers. Why do you wrong each other?' But the man who was wronging his neighbor*

thrust him aside, saying, 'Who made you a ruler and a judge over us? Do you want to kill me as you killed the Egyptian yesterday?' At this retort Moses fled and became an exile in the land of Midian, where he became the father of two sons. "Now when forty years had passed, an angel appeared to him in the wilderness of Mount Sinai, in a flame of fire in a bush. When Moses saw it, he was amazed at the sight, and as he drew near to look, there came the voice of the Lord (Acts 7:23–36).

Chapter Six

1. *The hand of the* LORD *was upon me, and he brought me out in the Spirit of the* LORD *and set me down in the middle of the valley; it was full of bones. And he led me around among them, and behold, there were very many on the surface of the valley, and behold, they were very dry. And he said to me, "Son of man, can these bones live?" And I answered, "O Lord* GOD, *you know." Then he said to me, "Prophesy over these bones, and say to them, O dry bones, hear the word of the* LORD. *Thus says the Lord* GOD *to these bones: Behold, I will cause breath to enter you, and you shall live. And I will lay sinews upon you, and will cause flesh to come upon you, and cover you with skin, and put breath in you, and you shall live, and you shall know that I am the* LORD."

So I prophesied as I was commanded. And as I prophesied, there was a sound, and behold, a rattling, and the bones came together, bone to its bone. And I looked, and behold, there were sinews on them, and flesh had come upon them, and skin had covered them. But there was no breath in them. Then he said to me, "Prophesy to the breath; prophesy, son of man, and say to the breath, Thus says the Lord GOD: *Come from the four winds, O breath, and breathe on these slain, that they may live." So I prophesied as he commanded me, and the breath came into them, and they lived and stood on their feet, an exceedingly great army.*

Then he said to me, "Son of man, these bones are the whole house of Israel. Behold, they say, 'Our bones are dried up, and our hope is lost; we are indeed cut off.' Therefore prophesy, and say to them, Thus says the Lord GOD: Behold, I will open your graves and raise you from your graves, O my people. And I will bring you into the land of Israel. And you shall know that I am the LORD, when I open your graves, and raise you from your graves, O my people. And I will put my Spirit within you, and you shall live, and I will place you in your own land. Then you shall know that I am the LORD; I have spoken, and I will do it, declares the LORD."

The word of the Lord came to me: "Son of man, take a stick and write on it, 'For Judah, and the people of Israel associated with him'; then take another stick and write on it, 'For Joseph (the stick of Ephraim) and all the house of Israel associated with him.' And join them one to another into one stick, that they may become one in your hand. And when your people say to you, 'Will you not tell us what you mean by these?' say to them, Thus says the Lord GOD: Behold, I am about to take the stick of Joseph (that is in the hand of Ephraim) and the tribes of Israel associated with him. And I will join with it the stick of Judah, and make them one stick, that they may be one in my hand. When the sticks on which you write are in your hand before their eyes, then say to them, Thus says the Lord GOD: Behold, I will take the people of Israel from the nations among which they have gone, and will gather them from all around, and bring them to their own land. And I will make them one nation in the land, on the mountains of Israel. And one king shall be king over them all, and they shall be no longer two nations, and no longer divided into two kingdoms. They shall not defile themselves anymore with their idols and their detestable things, or with any of their transgressions. But I will save them from all the backslidings in which they have sinned, and will cleanse them; and they shall be my people, and I will be their God.

"My servant David shall be king over them, and they shall all have one shepherd. They shall walk in my rules and be careful to obey my statutes. They shall dwell in the land that I gave to my servant Benjamin, where your fathers lived. They and their children and their children's children shall dwell there forever, and David my servant shall be their prince forever. I will make a covenant of peace with them. It shall be an everlasting covenant with them. And I will set them in their land and multiply them, and will set my sanctuary in their midst forevermore. My dwelling place shall be with them, and I will be their God, and they shall be my people. Then the nations will know that I am the LORD who sanctifies Israel, when my sanctuary is in their midst forevermore (Ezek. 37:1–28).

Chapter Nine

1. *The LORD said to Moses, "Tell Aaron and his sons, 'This is how you are bless the Israelites. Say to them: """The LORD bless you and keep you. The LORD make his face shine on you and be gracious to you. The LORD turn his face toward you and give you peace."""' So they will put my name on the Israelites, and I will bless them"* (Num. 6:22–27 NIV).

2. *Trust in the LORD with all your heart and lean not on your own understanding. In all your ways acknowledge him, and he will make straight your paths* (Prov. 3:5–6 NIV).

3. *Now the LORD said to Abram, "Go from your country and your kindred and your father's house to the land that I will show you and I will make of you a great nation, and I will bless you and make your name great, so that you will be a blessing. I will bless those who bless you, and him who dishonors you I will curse, and in you all the families of the earth shall be blessed"* (Gen. 12:1–3).

Chapter Ten

1. *The LORD spoke to Moses, saying, "Speak to the people of Israel and say to them, These are the appointed feasts of the LORD that you shall proclaim as holy convocations; they are my appointed feasts.*

   *[**The Sabbath**] "Six days shall work be done, but on the seventh day is a Sabbath of solemn rest, a holy convocation. You shall do no work. It is a Sabbath to the LORD in all your dwelling places.*

   *[**The Passover**] "These are the appointed feasts of the LORD, the holy convocations, which you shall proclaim at the time appointed for them. In the first month, on the fourteenth day of the month at twilight, is the LORD's Passover.*

   *[**The Feast of Unleavened Bread**] "And on the fifteenth day of the same month is the Feast of Unleavened Bread to the LORD; for seven days you shall eat unleavened bread. On the first day you shall have a holy convocation; you shall not do any ordinary work. But you shall present a food offering to the LORD for seven days. On the seventh day is a holy convocation; you shall not do any ordinary work."*

   *[**The Feast of Firstfruits**] And the LORD spoke to Moses, saying, "Speak to the people of Israel and say to them, When you come into the land that I give you and reap its harvest, you shall bring the sheaf of the firstfruits of your harvest to the priest, and he shall wave the sheaf before the LORD, so that you may be accepted. On the day after the Sabbath the priest shall wave it. And on the day when you wave the sheaf, you shall offer a male lamb a year old without blemish as a burnt offering to the LORD. And the grain offering with it shall be two tenths of an ephah of fine flour mixed with oil, a food offering to the LORD with a pleasing aroma, and the drink offering with it shall be of wine, a fourth of a hin. And you shall eat neither bread nor grain parched or fresh until this same day, until you have brought the offering of your God: it is a statute forever throughout your generations in all your dwellings.*

[The Feast of Weeks] "*You shall count seven full weeks from the day after the Sabbath, from the day that you brought the sheaf of the wave offering. You shall count fifty days to the day after the seventh Sabbath. Then you shall present a grain offering of new grain to the LORD. You shall bring from your dwelling places two loaves of bread to be waved, made of two tenths of an ephah. They shall be of fine flour, and they shall be baked with leaven, as firstfruits to the LORD. And you shall present with the bread seven lambs a year old without blemish, and one bull from the herd and two rams. They shall be a burnt offering to the LORD, with their grain offering and their drink offerings, a food offering with a pleasing aroma to the LORD. And you shall offer one male goat for a sin offering, and two male lambs a year old as a sacrifice of peace offerings. And the priest shall wave them with the bread of the firstfruits as a wave offering before the Lord, with the two lambs. They shall be holy to the LORD for the priest. And you shall make a proclamation on the same day. You shall hold a holy convocation. You shall not do any ordinary work. It is a statute forever in all your dwelling places throughout your generations. And when you reap the harvest of your land, you shall not reap your field right up to its edge, nor shall you gather the gleanings after your harvest. You shall leave them for the poor and for the sojourner: I am the LORD your God.*"

[The Feast of Trumpets] And the LORD spoke to Moses, saying, "*Speak to the people of Israel, saying, In the seventh month, on the first day of the month, you shall observe a day of solemn rest, a memorial proclaimed with blast of trumpets, a holy convocation. You shall not do any ordinary work, and you shall present a food offering to the LORD.*"

[The Day of Atonement] And the Lord spoke to Moses, saying, "*Now on the tenth day of this seventh month is the Day of Atonement. It shall be for you a time of holy convocation, and you shall afflict yourselves and present a food offering to the LORD. And you shall not do any work on that very day, for it is a Day of Atonement, to make atonement for*

you before the LORD *your God. For whoever is not afflicted on that very day shall be cut off from his people. And whoever does any work on that very day, that person I will destroy from among his people. You shall not do any work. It is a statute forever throughout your generations in all your dwelling places. It shall be to you a Sabbath of solemn rest, and you shall afflict yourselves. On the ninth day of the month beginning at evening, from evening to evening shall you keep your Sabbath."*

[The Feast of Booths] And the LORD *spoke to Moses, saying, "Speak to the people of Israel, saying, On the fifteenth day of this seventh month and for seven days is the Feast of Booths to the* LORD. *On the first day shall be a holy convocation; you shall not do any ordinary work. For seven days you shall present food offerings to the* LORD. *On the eighth day you shall hold a holy convocation and present a food offering to the* LORD. *It is a solemn assembly; you shall not do any ordinary work.*

"These are the appointed feasts of the LORD, *which you shall proclaim as times of holy convocation, for presenting to the* LORD *food offerings, burnt offerings and grain offerings, sacrifices and drink offerings, each on its proper day, besides the* LORD'S *Sabbaths and besides your gifts and besides all your vow offerings and besides all your freewill offerings, which you give to the* LORD. *On the fifteenth day of the seventh month, when you have gathered in the produce of the land, you shall celebrate the feast of the* LORD *seven days. On the first day shall be a solemn rest, and on the eighth day shall be a solemn rest. And you shall take on the first day the fruit of splendid trees, branches of palm trees and boughs of leafy trees and willows of the brook, and you shall rejoice before the* LORD *your God seven days. You shall celebrate it as a feast to the* LORD *for seven days in the year. It is a statute forever throughout your generations; you shall celebrate it in the seventh month. You shall dwell in booths for seven days. All native Israelites shall dwell in booths, that your generations may know that I made the*

people of Israel dwell in booths when I brought them out of the land of Egypt: I am the LORD your God." Thus Moses declared to the people of Israel the appointed feasts of the LORD (Lev. 23:1–44).

Chapter Thirteen

1. *I shall not die, but I shall live and recount the deeds of the LORD* (Ps. 118:17).

Chapter Seventeen

1. *Oh, taste and see that the LORD is good! Blessed is the man who takes refuge in him* (Ps. 34:8)!

About the Author

Audrey Lero is a world traveler who enjoys exploring cultures and meeting the people who create them. Israel, and its vineyards, is one of her favorite destinations. Using her gift of connecting people to the resources they need, she bridges cultural divides, inviting people to look past their differences to serve a common goal. Audrey is a graduate of Texas A&M University where she studied accounting and finance. She works professionally in real estate as both an agent and investor. Audrey enjoys writing from her Dallas, Texas home. She also writes and performs sketch comedy while spending as much time as possible outdoors.